NORTHWEST
GREENBOOK

NORTHWEST
GREENBOOK

A Regional Guide to Protecting and Preserving Our Environment

Jonathan King

Sasquatch Books
Seattle, Washington

Copyright ©1991 by Jonathan King
All rights reserved. No parts of this book may be reproduced without written permission of the publisher.
Printed in the United States of America.

Library of Congress Cataloging-in-Publication Data
King, Jonathan, 1956—
The northwest greenbook: a regional guide
to protecting and preserving our
environment/Jonathan King.
p. cm.
ISBN 0-912365-41-2: $9.95
1. Environmental protection—Northwest, Pacific.
2. Human ecology—Northwest, Pacific. I. Title.
TD 171.3.N67K56 1991
363.7'009795—dc20
91-10604
CIP

Printed on recycled paper ♻

Design by Karen Schober
Cover illustration by Don Baker

Sasquatch Books
1931 Second Avenue
Seattle, Washington
(206) 441-5555

Contents

Introduction
7

Growth
13

Timber
39

Water
79

Air
127

Energy
155

Recycling
167

Green Products
187

Appendixes
201

INTRODUCTION

Even in the most urban Northwest setting, natural beauty is never distant. The snowcapped summit of Mount Hood is visible from Portland; Seattle's waterfront opens up on Puget Sound and stunning vistas of the Olympic Peninsula, and on a clear day, Mount Rainier is sharply etched against the skyline.

But look more closely. There's a layer of brown haze hanging in the air. When you hit the freeway, the traffic is bumper-to-bumper. Once you shake free of the congestion and head east toward the Cascades, housing developments, shopping malls, and office complexes mar the landscape. Even up in the mountains, nature is not at peace: the hillsides are denuded; large swaths of trees have been mowed down. You don't have to look much beyond the picture-postcard views of the Northwest to notice that all is not right with the region's environment.

Equally serious are problems we cannot so readily see. Salmon runs on the Columbia and Snake rivers have declined precipitously, and many species are in danger of becoming extinct. Hundreds of toxic waste sites dot the landscape from Montana to Oregon. Even our abundant water resources are beginning to appear tapped out, stretched to the limit by proliferating, competing demands. Why is this happening to this once-pristine corner of the country, known for its vast forests, wide-open spaces, majestic mountains, and mighty rivers?

The obvious culprit is growth. A burgeoning population—particularly in Washington, where the number of residents grew by more than 730,000 in the 1980s—has put the squeeze on the region's natural resources. But blaming our environmental woes on population growth is too simple; the problems stem more specifically from *how* we've grown. For years, we have exploited our natural resources with little regard for future availability, ignoring warning signs of environmental stress. Despite dwindling supplies of timber, we continue to cut our forests at an unsustainable pace. Sprawl has eaten up land—wetlands, forestland, and farms—at a much faster rate than that at which the population is growing. Most of us are driving more, adding to air pollution; in the Puget Sound area, the number of vehicle miles traveled is increasing four times as fast as the number of people.

Clearly, the efforts of the last two decades to protect the Northwest's environment are not proving sufficient. Over the past 20 years, we have come to rely on regulation as our primary safeguard. This approach, in which government sets allowable levels of pollution and tries to force compliance through fines and other punishments, has enjoyed some success. It has tackled obvious sources of contamination, such as large factories that spew noxious smoke from huge stacks or pump industrial wastes into rivers and streams. But environmental agencies, never particularly well funded, have often been overwhelmed by the tasks confronting them. At times, political influence has played an undue role in environmental decision making. All of these factors have contributed to gaps in regulatory protection, as well as periodic compliance problems. There is little question that these agencies would benefit from additional resources and funds, particularly as regulations become more complex and responsibilities multiply.

But simply beefing up regulatory agencies and ratcheting down on large industrial sources of pollution aren't going to solve all of our environmental problems. The Northwest, with the exception of a few areas, is not heavily industrialized. Much of the pollution in the four states covered in this book—Washington, Oregon, Idaho, and Montana—stems from millions of dispersed smaller sources, such as motor vehicles, wood stoves, home septic fields, farms, and small businesses. In Washington, the most industrialized of the four, industry accounts for 25% of the air pollution. Wood stoves are responsible for almost as much—20% of the pollutants. The chief offenders are motor vehicles, the source of 40% of all air pollution. Much of the pollution running into our streams and lakes comes from agriculture and other land-use practices. Some regulation of these more dispersed sources is already in place, and more is in the offing. The city of Missoula, Montana, for example, has banned installation of fireplaces and severely restricted the use of wood stoves. King County, Washington, is considering banning wood stoves in new construction.

But it's virtually impossible for already overburdened regulators to clamp down on smaller-scale offenders: there are too many of them, their polluting is often intermittent, and they are difficult to monitor. It's laughable, not to mention undesirable, to picture an air-pollution control officer going from house to house to make sure that no one fires up the wood stove during a burn ban or that nobody takes an unnecessary Sunday drive.

Because of regulation's shortcomings, governments are now turning to financial incentives to encourage more environmentally sound behavior. Oregon is considering imposing fees on a wide range of air-polluting activities, such as field and wood burning. Seattle's solid waste utility uses a variable rate structure to promote recycling. Still, economic incentives and disincentives have limits. Higher fees are usually politically unpopular and often end up unfairly burdening lower-income people.

What this means is that northwesterners are going to have to voluntarily comply if the antipollution effort is to be truly successful. There's a growing realization that we all contribute to our environmental problems and to the strain on natural resources. The popularity of recycling programs and "green" products shows that the public wants to do something, and people are discovering that there are plenty of relatively painless life-style changes they can make, such as recycling and reusing, buying energy-efficient appliances, and keeping automobile engines tuned and tires inflated.

But individual actions are not by themselves going to win the battle against environmental degradation. Growth management, saving the salmon, preserving ancient forests, and avoiding destruction of the ozone layer are all too complex and too interrelated, and create too much confrontation, to yield to simple individual actions. The current wave of green advertising seems to imply that we can shop our way out of environmental problems, but the promotion of individual action must be seen for what it is—a useful educational tool. The crucial question is whether individual concern for the environment can be a catalyst for broader citizen involvement in environmental reforms. The organizations and resources listed at the end of each chapter in this book are included so that readers can become more involved in various arenas of environmental protection. These groups seek different objectives and employ different methods and tactics, but they are generally united in trying to protect and preserve the region's natural resources

The history of the environmental movement in the Northwest is filled with examples of citizen groups tackling tough environmental issues, such as the dwindling salmon runs on the Columbia or radioactive contamination from the Hanford nuclear weapons facility in Washington. These are generally tough, protracted struggles that involve sacrificing time, wrestling with government bureaucracies and entrenched economic interests, and negotiating compromises. The reward comes in seeing that citizens can make a clear contribution to the overall effort.

The 1990s will be a crucial decade for the Northwest's environment. Putting off difficult decisions on the fate of wildlife, forests, air, open space, and water could mean losing those precious natural resources for all time. There are not a lot of easy — or inexpensive—solutions. Achieving cleaner air will require more regulations and cost more money. Preserving old-growth forests will mean the loss of loggers' jobs. Saving water to aid salmon runs may mean higher electrical rates and less water for irrigation. But it is wrong to see preserving our environment as a sacrifice. Safe-

> *I ask the New York environmentalist about estuaries: are there any fish or migrating birds across the way where the Hudson River empties into the Atlantic? I tell him about the Bowerman Basin; the very week I'm in New York, Congress votes to set aside part of the mudflats in Grays Harbor for the sandpipers. He takes off his glasses and gives me a quizzical look, blinking. What do you mean? I ask if the Hudson and Atlantic convergence fosters any marine life. He shakes his head and says you wouldn't want to eat anything that came out of the water below, then points to the bank that rises up from the water, New Jersey, and the narrow band of mud at the foot of the cliff.*
>
> *"The estuary, you mean? Is that where it is?"*
>
> *"I guess," he says. "I'm not sure what you call it—I'm not familiar with the terms—but isn't it pretty?" I tell him not to apologize; some words leave the language when they no longer have any use.*
>
> —Timothy Egan
> The Good Rain, 1990

guarding the richness, diversity, and uniqueness of the Northwest's forests, rivers, and mountains is worth the cost and the effort. We have seen what a crippled environment can do to the economic—and human—health of a region. Eastern Europe and, closer to home, southern California provide stark examples of the costs of ignoring the environment. Many northwesterners were drawn here by the region's diverse resources and natural beauty. To give up on the fight to safeguard our environment would be tantamount to abandoning the region's soul, and our own well-being. This book is written in the hopes that we will not fail to act.

A final note: The effects of environmental damage are felt unevenly. Not surprisingly, pollution follows human beings; in more densely populated and industrialized areas—such as Puget Sound and metropolitan Portland—the effects of air pollution, water contamination, and suburban sprawl are relatively more pronounced. So, while this is a book about the environment of the entire Northwest, several chapters focus primarily on Washington and Oregon.

GROWTH AND LAND MANAGEMENT

LAND USE

GROWTH CONTROL: THE STATES STEP IN

TRANSPORTATION

OVERCOMING SELF-INTEREST

GROWTH AND LAND MANAGEMENT

If we were to look for one factor underlying most of the Northwest's environmental problems, the search would lead to growth—both economic and population growth. Growth drives the development that is burying our wetlands, farms, and forests under a layer of office buildings, parking lots, shopping malls, and housing developments. It's clogging our freeways, erasing the gains made in the fight against air pollution, reducing wildlife habitat, and straining our natural resources.

The stress on the region's environment stems from the *way* we have grown, not simply from growth itself. We have dammed and drained our rivers to feed often wasteful irrigation schemes and to provide cut-rate electricity to power-hungry industries such as aluminum smelting. We are producing ever-higher mountains of trash. And to make matters worse, development is chewing up open space at a much faster rate than the population is growing, exacerbating water pollution and flooding, and causing an increasing reliance on motor vehicles, already the leading cause of air pollution in many parts of the region.

If all goes according to projections, certain areas of the Northwest can expect little respite from growth and development. The combined population of western Washington's King, Kitsap, Snohomish, and Pierce counties, which increased by more than 20% in the last decade, is projected to grow by another *1.4 million*—roughly 50%—by 2020. That's the equivalent of moving metropolitan Portland into the Puget Sound area. The results of that kind of growth are predictable. Sprawl will climb up the sides of the Cascades and spill farther north and south along Interstate 5. There will be more garbage, more congestion, less open space, and greater demands on our natural resources.

Growth and its effects are felt unevenly in the Northwest. The entire state of Idaho has about a million people, and Montana has a population of only 800,000. By comparison, the Seattle-Tacoma area has about 2.6 million residents, and the Portland area has 1.5 million. Even within states, growth is extremely unbalanced. While Seattle booms, the economically devastated timber communities of the Olympic Peninsula are begging for development. The unevenness of economic development and population growth makes it extremely difficult to reach any kind of compromise on effective growth management: what booming areas see as necessary to alleviate the pressures of growth, economically depressed locales see as obstacles to attracting development.

One thing is certain: once an area is developed, it stays that way. You can't put sprawl back into the bottle. That's why, at least for some parts of the region, the clock is ticking when it comes to finding the means to manage growth. The next few years will tell whether we can preserve what is unique about the Northwest, or will instead continue to fail to learn from others' experiences and become, like

> *Set down there not knowing it was Seattle, I could not have told where I was. Everywhere frantic growth, a carcinomatous growth. Bulldozers rolled up the green forests and heaped the resulting trash for burning. The torn white lumber from concrete forms was piled beside gray walls. I wonder why progress looks so much like destruction.*
>
> —JOHN STEINBECK
> *TRAVELS WITH CHARLEY*, 1962

16 Northwest Greenbook

southern California, mile upon mile of undifferentiated development lying under a blanket of smog. There is, of course, no law etched in stone that growth must continue on its current path. The objective of growth management is not to stop development but to come up with plans to manage it so that it is more compatible with the environment.

LAND USE

Growth-related problems are often the result of a complex interplay of economic and social factors. The combination seldom bends to simple solutions. The challenge of controlling suburban sprawl, considered a key to reining in growth, illustrates this difficulty. On one level, the solution to sprawl is fairly obvious: contain it and encourage higher-density development. Environmentalists, planners, and politicians have all embraced higher-density development as necessary to control sprawl, make public transportation more efficient, preserve open space, and provide more affordable housing.

But there are precious few models of successful promotion of higher densities, and the Northwest is no exception. In central Puget Sound, for example, the number of acres developed has increased twice as rapidly as the population in the past 20 years, according to the Washington State Department of Ecology. That appetite for land is not confined to western Washington. Development in Spokane County has been chewing up acreage at a rate similarly disproportionate to population increase. One fundamental problem is the continued reliance on outdated, ineffective ways of managing land use.

The United States has traditionally relied on zoning as a way to channel development. But zoning,

> One fundamental problem is the continued reliance on outdated, ineffective ways of managing land use.

used to designate or limit the type of development that can take place on a specific tract of land, was born in an earlier time, when there were fewer people and much more undeveloped land. It doesn't work nearly as well today, when more people are crowding into less and less open space. For starters, the zoning system makes allowances for exemption or alteration. As James Ellis, a leading regional planner in the Seattle area, points out, "Significant changes in zoning can occur in deceptively small steps: large farms become 5-acre ranches, then 1-acre suburban estates, and eventually it becomes difficult to pinpoint exactly when the battle for open space was lost." UW geography professor Richard Morrill calls zoning a "farce" subject to modification and the prevailing political winds. Morrill points out that certain types of zoning can also encourage inefficient use of land by spreading development over large areas, which serves to prevent preservation of significant tracts of open space.

One increasingly popular way to steer development into already existing communities is the "urban growth boundary," a perimeter around an urban area, inside of which development is encouraged and outside of which it is restricted. But Morrill argues that such boundaries can actually exacerbate the very problems of sprawl, transportation, and housing affordability they are supposed to corral. Their establishment immediately puts upward pressure on the value of the limited amount of undeveloped land inside the boundaries, he explains. This, in turn, drives up home prices and encourages speculation as owners keep land off the market in hopes of continuing appreciation in value. Skyrocketing land prices inside the boundaries, in turn, lead to "leapfrogging," in which development skips the designated rural areas and pushes farther out into neighboring counties or other jurisdictions where land prices are much cheaper.

Given doubts about the effectiveness of urban growth boundaries, there is growing enthusiasm for the development of large-scale planned communities with higher density and prescribed boundaries. Ideally, these developments would include parks, recreational lands, and lower-income housing. Morrill argues that such communities, if planned correctly and buffered by open space, are an effective way to avoid the endless half-acre lots and cul-de-sacs that have characterized so much of the past few decades' development.

> The problem is precisely how to keep that land away from the developers' bulldozers.

Planners may be singing the praises of large-scale developments, but they aren't necessarily so popular with local residents who want to preserve the character of their rural neighborhoods. Often the alternative in fast-growing areas is a steady whittling-away of available land by numerous smaller developments. In the long run, these endless subdivisions destroy countryside to an even greater extent than do planned communities.

An essential part of maintaining the livability of rapidly developing areas is setting aside open space —whether agricultural, parks, wildlife refuges, or simply rural countryside. The problem is precisely how to keep that land away from the developers' bulldozers. Public purchase of open space is the one really proven means of preserving green space. "In urbanizing regions, only public purchase can permanently hold threatened farms, forests, sensitive wetlands, wildlife habitat, or outdoor recreation lands in open uses," James Ellis contends. Still, open space—like most things in life—costs money. Some local and state governments in the region have been active in purchasing open space. More than a decade ago, voters in King County approved a farmland preservation ordinance, and in 1989 they passed a $117 million bond issue to purchase greenbelts. The county commissioners of Skagit County, Washington, have allocated funds to help a local organization called Skagitonians to Preserve Farmland come up with recommendations to do just that. Skagit county residents have already successfully halted plans to construct an agricultural theme park on choice cropland.

Washington State has also put aside funds to purchase open space. During the 1989 legislative session, $53 million in state obligation bonds was authorized to purchase wildlife habitat and land for recreation. A major drawback to relying on public largess, however, is that during economic downturns and budgetary crunches, the enthusiasm for allocating scarce funds to buy up open space may wane.

Private organizations such as the Nature Conservancy and the Trust for Public Land also play an important role in preserving sensitive lands and open space. Since the early 1970s, for example, the Trust for Public Land has protected more than 23,000 acres in Alaska and the four northwestern states.

GROWTH CONTROL: THE STATES STEP IN

In the last few years, state governments have begun to take over some of the land-use planning that was formerly the bailiwick of local governments. This trend is due in large part to the perception that local government jurisdiction is too splintered and too subject to developer pressure to plan growth management effectively. States are the only entities with the power to manage growth comprehensively, says land-planning guru Ellis.

In the early 1970s, Oregon became one of the first states in the nation to institute a statewide land-use program, developed by the Land Conservation and Development Commission. Among the commission's goals are the preservation of agricultural lands and forests and the establishment of urban growth boundaries in the state's metropolitan areas. Local zoning and land-use regulations are required to be in line with these state goals. Oregon's state planning process has come under intense criticism at times, and three different initiatives have sought to repeal the law. But all were defeated.

But growth management, while more evolved in Oregon than its northern neighbor, is not entirely successful. A study by the state Department of Land Conservation and Development found that much of the development in Oregon's fastest-growing areas was occurring outside the urban growth boundaries. More than half the development in the Bend area, for example, occurred outside the boundary in the latter half of the 1980s. In addition, a lot of development in Oregon is still taking place in farm- and forestlands. "There is substantial evidence pointing toward a pattern and practice of systematic violation of the land use laws," claims Henry Richmond, executive director of 1000 Friends of Oregon, a land-use watchdog organization.

State land-use planning faces much more of an acid test in Washington, where the population is larger and growth has occurred rapidly in the counties bordering on Puget Sound. The Washington legislature passed a growth-management measure in 1990 calling for fast-growing counties to adopt comprehensive plans that include urban growth boundaries. It also ▷

WETLANDS

Until recently, wetlands were much maligned as bogs, swamps, or big mud puddles. They were consequently filled in, plowed over, paved over, and built on. It is estimated that in the 17th century, the lower 48 states contained over 200 million acres of wetlands. Now more than half of that is gone. Washington had 1.5 million acres of wetlands in 1850. Today 938,000 acres remain, and the state continues to lose between 700 and 2,000 acres each year. Idaho has lost more than half its original wetlands, while Oregon has seen 38% disappear. Montana has done comparatively well, losing only about a quarter of its natural wetlands.

Wetlands are lands that are inundated or saturated by water long enough to "support a prevalence of vegetation typically adapted for life in saturated soil," according to the federal government's definition. This definition covers a wide variety of environments, including coastal marshes, swamps, bogs, river deltas, and riparian areas (land along rivers or lakes). The Alaskan tundra is also considered wetland.

We discovered only belatedly just how essential these areas are. Wetlands improve water quality by filtering and reducing sediment, nutrients, and organic and chemical wastes. They act as a kind of natural sponge that soaks up excess water and thus helps reduce flood and storm damage. They help protect against erosion and provide vital breeding sites and sources of food for a wide variety of fish and wildlife, including many endangered species. "Acre for acre, wetlands match the most productive ecosystems in the world," the EPA states.

Currently, wetlands are supposed to be partly protected under the 1972 Clean Water Act, which gives the U.S. Army Corps of Engineers the power to issue permits for "the discharge of dredged or fill material into the navigable waters" of the United States. The corps also has primary responsibility for taking action against permit violators. It doesn't have the final word on permitting, however; the EPA can stop alteration or use of wetlands if unacceptable environmental damage is expected to result. In practice, though, the EPA very rarely asserts this power.

The Corps of Engineers receives about 13,000 applications for wetlands projects each year. Slightly more than 3% of permit applications are denied, and another 14% are withdrawn by the applicants. Approximately one-third of the permits are approved with major alterations. Such control saves 50,000 acres of wetlands annually, according to the congressional Office of Technology Assessment.

Even though the EPA has adopted a goal of "no net loss" of wetlands, they continue to disappear nationally at the rate of about 300,000 acres a year. Generally, this attrition is occurring in small parcels and not in dramatic swipes, says Fletcher Shives, an EPA regional staffer in wetlands enforcement. Part of the reason is exemptions from the permitting process. The federal government allows modification of wetlands if filling will cause "the loss or substantial adverse modification of less than one acre" of so-called isolated wetlands, that is, those not connected to other wetlands, or wetlands adjacent to and above the headwaters of small streams.

State governments play a role as well in wet-

> lands maintenance, mainly through standards for water quality. Construction in wetlands must also conform to shoreline protection regulations and local building guidelines. But often, local and state regulations are not strong enough to prevent continuing loss. In Washington, attempts to develop a comprehensive state program to manage and protect wetlands have been floundering in the state legislature for the past several years, caught in a crossfire between environmentalists and developers.
>
> On the bright side, Fletcher Shives points out that individuals can and do play a major role in protecting wetlands. In fact, regulators, lacking large inspection staffs, depend heavily on citizen reports to catch permit violators and evaders. "Individuals can see something and report it and have a positive effect," Shives explains. "Wetlands is an area where we can take action quickly. With wetlands [enforcement] it's a yes or no decision."

▷ mandates that counties inventory farmlands, forestlands, and critical areas such as wetlands.

But many environmentalists in Washington felt the 1990 growth-management legislation didn't go far enough, because it didn't apply to all counties and didn't have the power to force compliance with the planning goals. They put a much more far-reaching initiative on the November 1990 ballot, but it suffered a crushing defeat at the hands of the state's voters.

That loss apparently helped derail efforts at effective growth control, at least in 1991, when some kind of effective management proposal was expected to emerge from the legislature. The House did pass a growth-control package that would require cities and counties to come up with long-range plans for urban growth areas, and that contained measures to protect environmentally sensitive areas. But in the Senate, Bob McCaslin, a Republican from Spokane County, managed to delay its passage by introducing his own growth-management bill, which effectively gutted the House bill of almost all its significant protection measures.

Comprehensive management may be a casualty of the uneven growth that Washington has experienced in recent years. Not surprisingly, most of the support for statewide control comes from western Washington lawmakers whose districts along the Interstate 5 corridor are struggling under the burden of mushrooming growth. The opposition is concentrated in largely rural and economically depressed districts who see growth-control measures as an impediment to the kind of development they are desperately seeking.

TRANSPORTATION

Poor or ineffective land management is also a major cause of traffic congestion, because land use directly affects traffic patterns and vehicle use. There's no argument that traffic is a problem in urban areas such as Seattle and Portland. In the Puget Sound area, congestion has become an obsession. The Greater Seattle Chamber of Commerce has called transportation its number one concern. *The Seattle Times* devotes a regular column to it. Local radio stations hawk their traffic reports as the reason to tune in.

According to one estimate, the average Seattle commuter loses the equivalent of 14 working days a year just sitting behind the wheel. And the situation is getting worse all the time. Rush hour is lengthening at both ends of the day. The average speed during peak traffic hours has declined by nearly 25% since 1987. One traffic researcher rated Seattle the sixth-most-congested urban area in the country. The Puget Sound Council of Governments projects that the number of miles driven on Puget Sound–area highways will increase by at least 70% in the next 30 years.

And congestion isn't the only traffic-related problem. Transportation is also the leading cause of air pollution in much of the region. In King County, more than 90% of the carbon monoxide comes from transportation sources.

Any effective campaign to unsnarl traffic will have to have several components, according to G. Scott Rutherford, director of the Washington State Transportation Center at the University of Washington. First, the transportation system itself must be improved. This means some road building, but far less than we have seen in the past few decades. Major road construction isn't a workable option anymore; there isn't enough land and it's too expensive. Beyond that, more roads simply encourage more automobile travel, which in the long run won't do much to relieve congestion. "Nobody thinks we'll build our way out of this," Rutherford says.

A more effective way to improve transportation is to build high-occupancy-vehicle (HOV) lanes to move buses and carpools past rush-hour congestion. Metro, the public transit agency in the greater Seattle area, has already embarked on a program to construct a system of HOV lanes. Many people also advocate a light-rail system, and voters in the Seattle area will probably soon be asked to approve a sales tax increase to cover the local contribution to the rail's multibillion-dollar price tag. Critics of a headlong rush into light rail question how many cars such a system would take off the roads. In the Portland area, where 4 million automobile trips are made each day, the rail system carries only a few thousand commuters daily—and many of those must drive to park-and-ride lots. It seems clear that a rail system would only be effective in already heavily traveled corridors and cannot be expected to serve as a cure-all for regional transportation woes.

Another necessary element, according to Rutherford, is the use of incentives and penalties to encourage people to reduce the amount of driving they do, particularly during rush hour. These can take several forms, including boosting parking costs, instituting tolls, raising gasoline taxes, and subsidizing public transportation. There is evidence that this approach may be effective. According to James Ellis, in the Seattle downtown area, where there is a high concentration of employment and parking is relatively expensive, buses carry 35% to 40% of commuter traffic. Across Lake Washington in Bellevue, which has a high concentration of employment but also a lot of free parking, that proportion drops to 10%, whereas in areas where employment is scattered and parking is free, such as industrial parks, bus ridership fades to 2%.

Such measures have at least helped reduce the proportion of employees who commute alone in some cases. There are examples of employers who have been notably successful in reducing worker reliance on cars, largely by subsidizing public transportation and limiting free employee parking. In Bellevue, whereas the overall proportion of workers who drive solo to work is about 80%, only 60% of Puget Power employees commute alone; among USWest employees, the rate is only 30%. The Washington State legislature is trying to encourage more of these programs. It has passed a law that, in counties with more than 150,000 people, requires employers with at least 100 workers to develop plans to reduce the number of commuter trips. The goal is to cut the number of employee commuter trips by 35% by 1999.

If the Bellevue experience is any guide, measures to reduce drive-alone commuters will be more effective in larger workplaces. But even that kind of reduction isn't easy to achieve. As one Bellevue City Planning Department official points out, "It's going to be more than a matter of putting up some rideshare leaflets on a bulletin board." Scott Rutherford warns that if we don't slow down the increase in vehicle use, "we will automatically implement one of the most effective demand-management strategies that exists today: congestion."

Without changing how land is developed, it will be hard to make a dent in congestion. Increasing residential densities and proximity to workplaces and services can help reduce driving and make public transportation systems more efficient. Even so, neither a better mass transit system nor increased density nor more expensive parking will empty our freeways. In cities such as Seattle and Portland, public transportation accounts for only 3% or 4% of all urban travel. As Richard Morrill and fellow UW professor David Hodge point out, even if transit ridership were to increase by 50%, the number of public-transit trips as a proportion of all trips would amount to only 6%. Even cities with extensive, well-used public transportation systems suffer from congestion.

In all likelihood, automobiles will remain our dominant form of transportation. The rapid increase in car use is due in large part to social and economic factors—the rise of two-breadwinner families whose wage earners work in different locations, the relative cheapness of owning cars, a lack of affordable housing near workplaces—that are not going to yield readily to a quick technological or planning fix.

OVERCOMING SELF-INTEREST

One of the biggest obstacles to effective growth control is simple self-interest, which finds some of its most troublesome expression in the notorious NIMBY (Not In My Backyard) syndrome. Everyone wants public services and infrastructure improvements, but we prefer that someone else pay for them. Everyone wants the social problems created by growth to be taken care of, but not next door. Try siting a rail station, jail, homeless shelter, low-income housing, or AIDS housing, and just watch neighborhood opposition suddenly materialize.

We all want to enjoy the benefits of open lands, uncongested freeways, and rural living without having to pay the price. We want to "drive our cars while others take the bus, . . . to zone someone else's land as open space so we can inexpensively preserve our view; to have free parking at work, but to drive there on free-flowing expressways," as James Ellis says. Seattle developer David Sabey put the dilemma in a nutshell: "Nobody wants sprawl, but nobody wants density either."

> "Nobody wants sprawl, but nobody wants density either."

While the NIMBY mindset has often been instrumental in forcing public officials and planners to come up with better ways to deal with problems, it has also limited our ability to plan on any meaningful scale. After all, how can you build a rail system if you can't site stations?

The irony in all of this is that there is no shortage of ideas on how to manage growth, but unfortunately, the proposed solutions generally involve increasing costs, setting limits on development, and sometimes modifying life-styles.

The Northwest will have to be farsighted enough to realize that the dark side of uncontrolled growth—traffic congestion, air pollution, lack of affordable housing near workplaces—also limits life-style choices and involves social and economic costs. As long as everyone is saying, "Do it in someone else's neighborhood," the chances are slim that we will be able to cope with the consequences of growth before we have lost too much valuable time and too many resources. If we can just accept that it is in everyone's interest to manage growth so as to minimize pollution and environmental damage, then northwesterners will have gone a long way toward preserving the quality of life that helped draw so many of them to the region in the first place.

RESOURCES

ADOPT-A-PARK
Seattle Parks and Recreation
Volunteer Park Cottage
Seattle, WA 98102
(206) 684-4558

Plans projects that enable Seattle neighborhoods to take part in the care of their neighborhood parks and green spaces. Welcomes volunteers—no offer of labor is too small.

ALLIANCE FOR THE WILD ROCKIES
Box 8731
Missoula, MT 59807
(406) 721-5420

A coalition of more than a hundred businesses and smaller conservation organizations working to secure wilderness designation for unprotected lands in the Rocky Mountain region. Promotes this cause through media awareness, petition drives, and networking projects. Currently putting forward a five-state proposal designed to protect 13 million acres of wilderness lands, 2 million acres of national parks, and 1,000 miles of wild and scenic rivers. A proposed wildland recovery program would remove unnecessary roads and reestablish fish-spawning runs.

FRIENDS OF THE EARTH
Northwest Office
4512 University Way NE
Seattle, WA 98105
(206) 633-1661

The only regional Friends office outside of Washington, DC, chooses issues independently of the national organization but works for change on the federal level. Working through Congress and federal agencies, FOE targets wetland protection, water conservation, energy policy, toxic contamination on military bases, and protection of coasts from offshore drilling. Water conservation efforts include participation in the Yakima water project and work on the campaign to free the Elwha River of dams. Lobbying is done only on the federal level. Recent successes include stopping the Navy from dumping contaminated waste into Puget Sound, and the establishment of a national wildlife refuge in Grays Harbor, WA. Membership organization.

FRIENDS OF THE SAN JUANS
PO Box 1344
Friday Harbor, WA 98250
(206)378-2319

The Friends' mission statement mandates the defense of comprehensive planning in the San Juan Islands and the protection of the islands' scenic, aesthetic, economic, sociological, and ecological assets. Organization focuses on public education and monitors planning commission meetings. A recent grant from the Puget Sound Water Quality Authority will fund public education on the proposed designation of northern Puget Sound as a national marine sanctuary.

FRIENDS OF UNION BAY
5026 22nd Avenue NE, Suite 2
Seattle, WA 98105
(206)525-0716

Organization working to protect the heavily development-targeted area of Seattle's Union Bay. Falling victim to a nearby municipal landfill and sewage sludge from all over the area, Union Bay was protected by the Friends against two major construction proposals in 1990: the extension of the Burke-Gilman Trail and a pipeline from the bay to Green Lake. Also works on educational projects with the State Superintendent of Environmental Education.

IDAHO WILDLIFE COUNCIL
PO Box 7043
Boise, ID 83708
(208)344-5159

Smaller affiliate organizations bring their issues to this council, which uses coalition power to carry out projects such as wilderness designation, wolf reintroduction, and the prevention of roads into proposed timber areas.

MONTANA ENVIRONMENTAL INFORMATION CENTER
PO Box 1184
Helena, MT 59624
(406)443-2520

Watchdog group spun off from the Clark Fork Coalition in 1973. Forces mining companies and agencies to comply with the Montana Environmental Policy Act and works on other legislation. Looks at hard-rock mining, solid and hazardous waste, and subdivision reform. Accomplishments include work on state environmental bills, the Montana Environmental Policy Act, the Subdivision and Platting Act, the Montana Pesticides Act, the

Resource Indemnity Act, and passage of the state Superfund law. Recently aided the residents of Mill Creek (western Montana) in acquiring funds from the EPA for relocation away from toxic mines.

MONTANA LAND RELIANCE
PO Box 355
Helena, MT 59624
(406) 443-7027

Private landowners donate development rights to their properties to this professionally staffed group, which in turn uses conservation easements and deed restrictions to protect the land.

NATIONAL AUDUBON SOCIETY
Rocky Mountain Regional Office
4150 Darley Avenue, Suite 5
Boulder, CO 80303
(303) 499-0219

This office, whose scope includes Montana, focuses its efforts on land-use and water quality issues. Involved in mapping out old-growth areas and pushing wilderness designation bills before state legislatures. Also working on reform of the permissive 1872 mining laws and water rights for wildlife. Regional offices act as links between national organization and local chapters.

Washington State Office
PO Box 462
Olympia, WA 98507
(206) 786-8020

Concerned with forestry on all levels—federal, state, and private—and lobbies on other environmental issues as well (oil spills, wetland development). Audubon chapters around the state address local issues.

Western Regional Office
555 Audubon Place
Sacramento, CA 95825
(916) 481-5332

This office, whose scope includes Oregon, acts as liaison between the New York office and local chapters. Addresses ancient-forest matters and other issues of development and growth.

NATIONAL WILDLIFE FEDERATION

Northern Rockies Natural Resources Center
240 N Higgins Street
Missoula, MT 59802
(406) 721-6705

Pacific Northwest Natural Resources Center
519 SW Third Avenue, Suite 606
Portland, OR 97204
(503) 222-1429

Both regional offices of the National Wildlife Federation address public land management and endangered species issues, using public education whenever possible and resorting to litigation if necessary. The Pacific Northwest office is garnering support for the Washington, Oregon, and Idaho Ancient Forest Protection Act; the Northern Rockies organization is working to block a proposed dam at Kootenay Falls. State chapters are separate from the national organization.

THE NATURE CONSERVANCY

Idaho Field Office
PO Box 64
Sun Valley, ID 83353
(208) 726-3007

Montana Field Office
PO Box 258
Helena, MT 59624
(406) 443-0303

Oregon Field Office
1205 NW 25th Avenue
Portland, OR 97210
(503) 228-9561

Washington Field Office
1601 Second Avenue, Suite 910
Seattle, WA 98101
(206) 343-4344

The conservancy's mission is the same all over the country. Uses a free-market approach to land protection, purchasing endangered lands and selling them to both public and private agencies who will manage and protect them. In Washington, this has been renamed the Endangered Washington Campaign, which garners matching state funds. Extensive use of volunteers, from office work to helping biologists in the field.

SIERRA CLUB

Cascade Chapter
1516 Melrose Street
Seattle, WA 98122
(206)621-1696

Involved in a range of issues on a federal level, from ancient forest protection to Wild and Scenic river designations to the Clean Air and Water acts. Also has an agenda in the state legislature.

Montana Chapter
c/o James Conner
78 Konley Drive
Kalispell, MT 59901
(406)752-8925 or contact the field office in Sheridan, WY, at (307)672-0425

The chapter's conservation chair has identified six areas of concern for 1991, ranging from wilderness and endangered species protection to energy and clean air issues. Methods include everything from education to legal action when necessary.

Northern Rockies Chapter
c/o Edwina Allen
1408 Joyce Street
Boise, ID 83706
(208)344-4565

Besides the most pressing Northwest issues—wilderness designation, ancient forests, endangered wild salmon, grass burning—this chapter is involved in Idaho's incipient recycling scene, pressuring legislators to look at the landfill issue and address the solid waste problem. Does a great deal of lobbying and electoral activities.

Oregon Chapter
c/o John Albrecht
3550 Willamette Street
Eugene, OR 97405
(503)343-5902

Concerned with a range of wilderness issues, including management of ancient forests in the Pacific Northwest. Also focuses on high desert wilderness, endangered species, and heap leach mining (which employs cyanide to extract gold). Involved in research and legal action to force compliance with clean air laws.

SIERRA CLUB LEGAL DEFENSE FUND

Northwest Office
216 First Avenue S, Suite 330
Seattle, WA 98104
(206)343-7340

Rocky Mountain Office
1631 Glenarm Place, Suite 300
Denver, CO 80202
(303)623-9466

Group of attorneys who work on environmental cases. Northwest emphases include litigation in forestry and wildlife issues; lawsuits include litigation against BLM for timber sales and against the U.S. Forest Service to obtain an injunction prohibiting logging in areas that might be critical habitat for the spotted owl. Legal proceedings also aimed at regulation of pulp mills discharging dioxins into Northwest waters. Seattle office is scheduled to move in fall 1991.

1000 FRIENDS OF OREGON
300 Willamette Building
534 SW Third Avenue
Portland, OR 97204
(503)223-4396

Serves as a private, nonprofit watchdog organization for Oregon land-use planning laws. 1000 Friends is active in public education on land planning issues and growth management. The organization is also active in the legal arena and runs a "Cooperating Attorneys Program," which joins citizens who want to take action against development with attorneys versed in land-use law. Funded by membership dues and donations.

THE TRUST FOR PUBLIC LAND
Northwest Regional Office
Smith Tower, 506 Second Avenue
Seattle, WA 98104
(206)587-2447

Operates in six states (the four in this book plus Alaska and Wyoming). Works with citizen groups and public agencies, protecting open space from development. Responsibility for these lands is eventually turned over to public agencies or private, nonprofit land trusts. Since its inception in 1973, 500,000 acres in 34 states have been protected; in the Northwest, the trust has purchased 23,371 acres valued at more than $41 million and has put more than 4,500 acres of the Columbia Gorge National Scenic Area in public ownership. Can put you in contact with local land trusts.

Urban Wildlife Coalition
935 Kirkland Avenue
Kirkland, WA 98033
(206) 622-5260

Focuses on promoting wildlife in urban areas within the Puget Sound region. Projects include backyard wildlife areas, distribution of signs designating nature trails, and waste reduction work. Some on-site evaluation, and hands-on work with animals. Volunteer opportunities available.

Volunteers for Outdoor Washington
4516 University Way NE
Seattle, WA 98105-4511
(206) 545-4868

Promotes "volunteer stewardship of Washington's outdoor recreation and natural resources." In addition to recruiting and coordinating volunteers, offers workshops in various aspects of trail maintenance. Volunteers specify interests and capabilities; staff members coordinate them with appropriate projects. Volunteer opportunities include an annual cleanup of Seattle's Green Lake, and shoreline maintenance and rehabilitation of the Green and Duwamish rivers.

Washington Environmental Council
4516 University Way NE
Seattle, WA 98105
(206) 527-1599

An umbrella group for about 90 of Washington State's environmental organizations. Focuses on educational work, but also has a legislative lobbying program in Olympia and actively monitors local agencies. Has been involved in wetland protection, the implementation of the state Superfund Law, and the expanding of the Timber, Fish, and Wildlife process. Volunteers monitor water quality and wildlife habitats.

Washington Native Plant Society
University of Washington/Botany, KB-15
Seattle, WA 98195
(206) 543-1682 or (206) 543-1976

The society's conservation committee looks at the impact of development and grazing on the state's greenery (for example, the scheduled broadening of Highway 410 near Mount Rainier) and counters with letter-writing campaigns and articles in support of lawsuits and direct action. Sponsors field trips and plant identification workshops.

WASHINGTON TRAILS ASSOCIATION
1305 Fourth Avenue, Suite 518
Seattle, WA 98101
(206) 625-1367

Main focus is maintenance and upkeep of trails. Issues center on land use such as clear-cutting and conflicts between different types of trail users (mountain bikers, hikers), but otherwise leaves environmental issues to environmental groups and sticks to hiker advocacy. Publishes *Signpost* magazine.

WASHINGTON WILDERNESS COALITION
PO Box 45187
Seattle, WA 98145-0187
(206) 633-1992

Works as a grassroots umbrella organization of small local groups on a range of issues (ancient forest preservation, public lands management, Wild and Scenic designations, wolf reintroduction, wilderness management, and climate change). Focus is on federal public lands, working through education, media relations, and coalition building.

WASHINGTON WILDLIFE AND RECREATION COALITION
112 Fourth Avenue E, Suite 202
Olympia, WA 98501-1191
(206) 754-1898

This single-issue organization (chaired by former U.S. senator Dan Evans, former U.S. representative Mike Lowry, and Governor Booth Gardner) aims to secure $500 million dollars in funding to acquire land for open space and recreation over the next decade. In 1990 the coalition acquired $53 million; the mid-1991 total was between $50 million and $60 million, with the possibility of a $250 million long-term commitment before the legislature. The money will be distributed to various agencies in the state, which in turn will buy land; funds are also available to cities and counties.

TIMBER

THE ROOTS OF THE OLD-GROWTH CRISIS

ENTER THE DISAPPEARING SPOTTED OWL

THE FOREST SERVICE'S BATTLE WITHIN

TIMBER CUTTING'S ENVIRONMENTAL LEGACY

AN ATTEMPT AT NEGOTIATION FALLS APART

WHITHER TFW?

TIMBER'S FUTURE

TIMBER

No single environmental issue has so polarized the Northwest as the fate of the region's forests. It has pitted environmentalists against the timber industry and often against federal and state regulators as well. Caught in the crossfire are timber workers—desperate to preserve their jobs, their communities, and their way of life—and a reclusive owl whose dwindling population is seen as a harbinger of the collapse of an ecosystem.

The two major areas of dispute concern the preservation of old-growth forests and whether private industry and state agencies are managing timber resources in a manner that is sensitive to the environment and that preserves wildlife habitat. Although the two debates often involve different players, they are linked by the fundamental problem of how a limited timber supply should be regulated—and for whose benefit.

The debates have raged in court, in the halls of Congress, in state legislatures and agencies, and in the media. They have been carried into the streets and into the woods. Radical environmentalists have chained themselves to logging equipment to save the giants of the Northwest forests from loggers' chain saws. Loggers have countered with public rallies against the "greenies." A gathering in 1990 attracted thousands of loggers and their families from all over the Northwest, who descended on Kelso, Washington, in a truck cavalcade 5 miles long. In early 1991, a northern spotted owl was killed and nailed to a sign in the Olympic National Park. Attached to the owl was a note: IF YOU THINK YOUR PARKS AND WILDERNESS DON'T HAVE ENOUGH OF THESE SUCKERS, PLANT THIS ONE. THEY TALK OF SO-

CIAL UNREST. THE MATCH HAS YET TO BE STRUCK. Although the act was condemned by loggers' organizations, it was indicative of how inflamed passions have become—and how high the stakes are.

For decades, timber formed the backbone of the Northwest's economy. But now, the region is facing the unpleasant reality that its towering forests of Douglas fir, Sitka spruce, western hemlock, and cedar, nurtured over hundreds of years in the wet climate west of the Cascades, have been logged to the point at which a major ecosystem is endangered. At the same time, scientists have come to understand what a rich, unique resource the Northwest's old-growth forests are. These venerable, centuries-old trees form "the most productive and massive forests that exist anywhere in the world," according to University of Washington forestry expert Jerry Franklin. They contain the largest bird populations of any coniferous forest on the continent, as well as more than 200 species of animals and fish. Some 1,500 species of invertebrates can exist in a single stand of old-growth trees.

Only a small fraction—perhaps 10%—remains of the forests that blanketed the western portions of Washington and Oregon when white settlers first arrived. Most of it has been logged in the past few decades. The Olympic Peninsula, for example, has lost a third of its old growth since 1974, according to the Wilderness Society. Very little old growth survives on private and state lands; the vast majority of what is left is located on federal lands managed by the U.S. Forest Service (USFS), the Bureau of Land Management (BLM), and the National Park Service. But even on federal property, these forests are threatened. Much of the federal timber is not protected from cutting; only trees in national parks and selected parts of national forests such as wilderness, national monuments, and research areas cannot be logged. If the rate of felling that occurred during the 1980s were to continue unabated, old-growth forests, aside from those in national parks and other protected areas, would disappear in a few decades.

The actual damage may be worse than the figures indicate. Industry's preferred method of logging old-growth timber is clear-cutting—leveling every tree on a given piece of land. On federal lands, clear-cuts, typically about 40 acres in size, are often interspersed with uncut stands of trees. The result resembles a patchwork quilt. This fragmentation of old growth into small, isolated stands can alter the temperature and humidity of the remaining groves, increase exposure of resident animals to predators, and cause increased wind damage, drastically diminishing the "biological value" of the remaining forests. In addition, logging has been more concentrated at lower elevations, where the trees are larger and more accessible. That choice makes sense economically—more timber per acre —but it has left most of the uncut old growth at higher elevations, where the forests are biologically less diverse. Research by the Wilderness Society shows that less than 20% of the remaining ancient forests are at low elevations.

> *The trail took us speedily into a forest-temple. Long years of labor by artists the most unconscious of their skill had been given to modelling these columnar firs. Unlike the pillars of human architecture, chipped and chiseled in bustling, dusty quarries, and hoisted to their site by sweat of brow and creak of pulley, these rose to fairest proportions by the life that was in them, and blossomed into foliated capitals three hundred feet overhead.*
>
> —THEODORE WINTHROP
> CANOE AND SADDLE, 1862

THE ROOTS OF THE OLD-GROWTH CRISIS

Although the forest crisis has claimed the public spotlight only in the last few years, its origins go back decades. During the early part of this century, timber was so cheap and plentiful in the Northwest that much of it was simply wasted. (The Forest Service found that during the early 1900s, more wood was wasted per acre in Washington than in any other area of the nation.) By World War II, however, supplies of timber on private lands were becoming depleted, and the timber industry turned to publicly held supplies. Federal timber sales in the Northwest started climbing in the 1940s and continued to increase dramatically for about 15 years. The annual take from national forests more than doubled between 1950 and 1965. Then, in the mid-1960s, the size of the federal harvest more or less leveled out.

Meanwhile, harvests from private timberlands were still declining. Because it takes about 50 years for trees such as Douglas firs to reach a height at which they are considered commercially harvestable, replanted lands would not be ready for harvest for decades. The handwriting was on the wall. In the early 1970s, a Forest Service report projected that timber production on timber-industry lands in western Washington and Oregon would drop sharply by the year 2000, and that employment in the wood and paper products industries would decline by 45% between 1970 and 2000.

The warning was largely ignored. Since the early 1970s, logging on federal lands has continued at a hot and heavy pace (with the exception of a sharp, temporary drop during the recession of the early 1980s). In fact, annual timber sales from federal

JOBS VS. ENVIRONMENT?

Since the dawn of the environmental movement and the birth of the Environmental Protection Agency (EPA) in 1970, a debate over jobs vs. environment has raged. The usual scenario involves a government attempt to clamp down on pollution, generating industry claims of massive job loss and economic disaster as the costs of complying with more stringent regulations.

Although environmental regulations have caused some job loss and do involve financial sacrifice, very rarely have industry's predictions of dire economic consequences come true. Back in 1971, for example, the petroleum industry estimated that it would cost $7 billion a year to gradually lower the level of lead in gasoline. The actual annual cost turned out to be between $150 million and $500 million. That's an overestimate of more than 1000%.

More recently, in the debate over the revision of the federal Clean Air Act, a report prepared for the Clean Air Working Group, an industry lobbying organization, predicted that the annual cost of the 1990 clean air amendments would total from $51 billion to $91 billion. By contrast, the EPA estimates that the costs of the final legislation will increase gradually to $25 billion a year by 2005. And those figures don't take into account the benefits of cleaner air.

Similar confusion swirls around the anticipated economic impact of the spotted owl restrictions in the Northwest. The federal government projects job losses at 28,000 (many of which, some observers feel, would have disappeared eventually due to dwindling timber supplies), while one tim-

Timber 45

> ber industry estimate predicts more than 100,000.
>
> In spite of numerous studies to the contrary, there is a continuing public perception that environmental regulations bring about massive job losses. When EPA staffers tried to investigate lead contamination at the Bunker Hill smelter in northern Idaho, they "encountered an extremely hostile attitude from the community," according to an EPA report. The community, which suffered from 50% unemployment, believed the smelter had been closed "because of the high cost of compliance with EPA regulations.... In the community's view, jobs were more important than lead contamination."

lands in western Washington and Oregon from 1977 to 1988 averaged about 15% higher than the levels called for in the Forest Service's management plans. That probably should have come as no surprise, since the Forest Service has traditionally seen the selling of timber as its primary mission. The agency has devoted more money to cutting down trees than to any other land-management option (such as recreation or wildlife management) and has subsidized timber cutting by building approximately 350,000 miles of logging roads nationwide over the past half century.

But the Forest Service wasn't acting on its own. It was under tremendous pressure to increase the rate of cutting, not only from the timber industry, but also from the service's own overseers in the U.S. Department of Agriculture and from members of the Northwest congressional delegation.

ENTER THE DISAPPEARING SPOTTED OWL

With old growth disappearing at such a rapid clip, environmentalists were desperate to stop wholesale logging on public lands. In the 1980s, they found what would prove to be a convenient tool: the northern spotted owl, a resident of the ancient forests from northern California through western Oregon and Washington into Canada. As forests were disappearing under loggers' saws, so was the spotted owl's natural habitat. Environmentalists and scientists seized upon the owl as an "indicator species" whose decline would warn of the destruction of the old-growth ecosystem. "The spotted owl is almost certainly just the tip of the iceberg," Jerry Franklin told a congressional panel in 1990. "There are probably dozens of other species as threatened as the owl."

In response to concern about the spotted owl's viability, the U.S. Fish and Wildlife Service (FWS) reviewed the status of the bird in 1982, but decided against listing it as a threatened species under the federal Endangered Species Act. (Under the act, a species is considered endangered when it is in peril of becoming extinct over all or a significant part of its range. A "threatened" designation means that a species is likely to become endangered in the near future.) Still, concern over the loss of owl habitat continued to grow, and in the mid-1980s the Washington Department of Wildlife listed the spotted owl as endangered; Oregon listed it as threatened.

Getting the federal government to officially recognize the problem proved more difficult. In 1987, conservationists petitioned Fish and Wildlife to list the owl as a threatened species, but the service declined. A legal appeal reversed that decision, and the

FWS was forced to undertake another review. (A congressional investigation later found that Fish and Wildlife's decision was not based on scientific research and had been largely politically motivated.) In the meantime, environmentalists went to court to try to slow down the rate of cutting, arguing that the Forest Service wasn't adequately protecting the owl.

While lawsuits and legal arguments over federal timber sales swirled in the late 1980s, the U.S. government itself took an action that promised to profoundly change the way federal forests in the Northwest are managed. Four federal agencies—the Forest Service, the Fish and Wildlife Service, the National Park Service, and the Bureau of Land Management—convened a panel of scientists under the leadership of a Forest Service biologist named Jack Ward Thomas to come up with a conservation strategy for the owl. The panel issued its plan in April 1990. The Interagency Scientific Committee (ISC) report proved to be a watershed event for logging in the Northwest. "We believe that the current situation . . . is unacceptable and has contributed to a high risk that spotted owls will be extirpated from significant portions of their range," the report concluded.

> *I say Oregon can—and will—have the old growth and the new growth. We can preserve the open spaces and natural areas, and still develop the infrastructure and business we need to maintain the very high quality of life Oregon needs.*
>
> —GOVERNOR BARBARA ROBERTS
> MARCH 26, 1991

The Thomas committee, as the panel was informally called, recommended that large segments of the remaining old-growth forest be placed off-limits to logging, and called for what amounted to a 25% reduction in the amount of Forest Service timber made available for harvest. BLM timberlands would be even harder hit; the bureau estimated that if the Thomas committee's recommendations were fully implemented, the yearly allowable sale of timber from more than 2 million acres of timberland the BLM manages in western Oregon would be reduced by about 60%. As drastic as the conservation strategy appeared, the scientists noted that in a "worst-case scenario," the already depleted owl population could still drop by half. The panel rather dryly predicted the storm its report would produce: "The immediate response, we expect, will be to focus almost solely on the short-term economic and social impacts of implementing the strategy as it affects availability of timber."

The outcry was indeed immediate. "Welcome to the demise of the Pacific Northwest," proclaimed Oregon congressman Bob Smith. Loggers declared that the plan condemned timber-dependent communities to economic purgatory, all for the sake of about 1,500 pairs of owls. "The recommendation looks like they're saying, 'Shut the entire Olympic National Forest down,' " the president of a Hoquiam, Washington, logging company told *The Seattle Times*. The American Forest Resource Alliance, a forest industry organization, estimated that implementation of the spotted owl conservation plan in Washington, Oregon, and northern California would result in the loss of about 100,000 timber-related jobs. The federal government put the total job loss by the year 2000 at a more modest, but still drastic, 28,000. Oregon timber workers, who are particularly dependent on federal supplies, would be especially hard hit.

Close on the heels of the Interagency Scientific Committee report, the timber industry received another blow. In June 1990, the Fish and Wildlife Service finally declared the spotted owl a threatened species. That meant more protection for the owl: harming or harassing spotted owls would be prohibited on private and state lands as well as on federal lands.

In the wake of the spotted owl's listing as a threatened species, the U.S. Interior Department (of which the Fish and Wildlife Service is part) started to put together a recovery scheme to ensure restoration of the bird's population. In February 1991, Interior Secretary Manuel Lujan appointed a team to come up with a plan, which is scheduled for release in 1992.

While there was plenty of hyperbole in the timber industry's claims of devastation, as well as demagoguery on the part of some politicians, the ISC report indeed signaled the end of an era for the Northwest's timber industry. "We are not going to have a glide path down. There's going to be a drop-off," observes Gary Cordova, a spokesman for USFS Region 6, which covers Oregon and Washington. Lower timber sales mean smaller Forest Service budgets and staff reductions. "We'll be lean and mean by fiscal year 1993," says Steve Paulson, who heads the Forest Service's timber sales program for the region. Planned timber sales on Forest Service lands in Oregon and Washington were reduced in fiscal 1991—to roughly 3 billion board feet from an average of about 4 billion in 1989 and 1990. (A board foot is equivalent to a piece of wood 1 foot square and 1 inch thick.) In the Olympic National Forest, planned timber sales have already dropped about 80% since 1987.

Paulson said in an interview in February 1991 that the Forest Service could put about 3 billion board feet up for sale annually in the 1990s if it adhered to the Thomas committee's recommendations. But it quickly became clear that even those levels wouldn't be achieved, because of delays and environmental concerns. By May 1991, the Forest Service said it could probably come up with only about two-thirds of the planned sales levels.

But even those reduced projections didn't consider the effects of continued court challenges by environmentalists, who used the federal government's continued hesitation in protecting the spotted owl to open a new round of legal assaults. The Forest Service tried to stake out a position that its management plans would not be inconsistent with the Thomas committee's recommendations. The BLM, which reduced its timber targets by about 30% in 1991, agreed not to log in the spotted owl conservation areas for a two-year interim period covering 1991 and 1992, while resource management plans are being prepared. However, the bureau wanted to reserve the right to allow logging in forests that provide "dispersal linkages" between owl habitat areas. These linkages, while not in the habitat areas designated by the ISC, enable the birds to migrate from one habitat area to another, and are crucial to the ability of young owls to find new habitat or repopulate stands where other owls have died.

The Forest Service was taken to court by environmentalists, who claimed that the agency had not taken proper steps to protect the owl. In March, U.S. District Judge William Dwyer ruled in their favor. In May, Dwyer issued an injunction giving the Forest Service until March 1992 to come up with a plan to protect the spotted owl and prevent it from selling any logging rights in owl habitat until the plan is completed. The failure to develop a plan, Dwyer said in a strongly worded opinion, "exemplifies a deliberate and systematic refusal by the Forest Service

Timber 51

and the FWS to comply with the laws protecting wildlife." The judge went on to lay the blame right on the desks of the executive branch: "This is not the doing of the scientists, foresters, rangers, and others at the working levels of these agencies. It reflects decisions made by higher authorities in the executive branch of government."

On top of that, the Fish and Wildlife Service announced a plan in April 1991 to declare more than 11.6 million acres as critical habitat for the owl. Logging could be restricted on those lands, which include 3 million privately owned acres. That's several million acres more than the Thomas committee had recommended for set-aside just a year earlier. (Logging is already prohibited on more than 3 million acres that are in national parks or wilderness areas.) "The Jack Ward Thomas strategy was a maintenance posture which assumed a 50% loss of the owl population," Fish and Wildlife Service director John Turner told a congressional panel. "I think our standard has to be higher under the Endangered Species Act." Again, the FWS action was the result of environmentalists taking legal action against the agency for failing to designate the owl's critical habitat when it was declared threatened.

Condemnation and dire predictions of economic devastation once again came from the timber industry and Northwest politicians. Oregon senator Mark Hatfield criticized the Fish and Wildlife plan as "biology run amok." And the president of an Oregon lumber company said, "They just created Appalachia in the Northwest."

The industry is clearly pinning its hopes on Secretary Lujan's spotted owl recovery strategy. In theory, the recovery plan is supposed to restore the owl population. But environmentalists are critical of the motivations of Lujan's panel, which they claim lacks scientific expertise. They are particularly suspicious of the secretary himself, who has not been muted in his displeasure over the Endangered Species Act and spotted owl protection measures. At one point, the Interior Secretary said, "Maybe we should change the law.... The spotted owl business is probably the prime example." Given Lujan's orientation, environmentalists almost take it as a given that the panel will allow political and economic considerations to triumph over biology when it issues its plan. "I'm skeptical [of the plan] until I see it," says Jean Durning of the Wilderness Society.

If the past pattern holds, whatever recommendations the recovery team comes up with won't do much to resolve the old-growth fight, particularly if the absence of leadership on the part of the federal government continues. If nothing else, the past several years of conflict over the fate of the ancient forests have shown how a lack of farsighted and decisive leadership can result in almost complete confusion. Politicians are constantly looking for short-term remedies that don't address the underlying problems. Agencies such as the Forest Service and the Bureau of Land Management are caught in their traditional roles as managers of timber harvests and are under intense political pressure from their overseers in both the executive branch and in Congress. The result is that they attempt compromises that end up pleasing no one and leaving them vulnerable to legal challenge.

Into this leadership vacuum have stepped the federal courts, which seem to be the only parties currently capable of forcing all the players to recognize that we have reached the point of no return. As long as the vacuum remains unfilled, the courts promise

> The past several years of conflict over the fate of the ancient forests have shown how a lack of farsighted and decisive leadership can result in almost complete confusion.

Timber 53

to be the only source of definitive decisions on the timber question.

THE FOREST SERVICE'S BATTLE WITHIN

While the Forest Service is being squeezed by outside forces, it is also being pressured from among its own ranks to transform its priorities. One of the major sparks for internal change was a U.S. Forest Service timber-sale planner in Oregon's Willamette National Forest named Jeff DeBonis. The Willamette National Forest carries the distinction of having more timber cut in it than any other national forest. By 1989, DeBonis had become so appalled at the rate of logging in national forests that he shot off a candid letter to Forest Service chief F. Dale Robertson. "We are overcutting our National Forests at the expense of other resource values. We are incurring negative, cumulative impacts to our watershed, fisheries, wildlife, and other noncommodity resources in our quest to meet our timber targets," he wrote. "We [the Forest Service] are perceived by much of the public as being dupes of, and mere spokespeople for, the resource-extraction industries." DeBonis spoke out publicly as well, incurring the wrath of the timber industry. At least one timber operator tried (unsuccessfully) to get the Forest Service to muzzle him.

DeBonis's efforts didn't stop there. In 1989, he founded an organization for Forest Service dissidents called the Association of Forest Service Employees for Environmental Ethics (AFSEEE). The outspoken forester became a media sensation, widely profiled and quoted in newspapers across the Northwest. His message clearly touched a rich vein of support among disgruntled Forest Service employees;

within a short time, AFSEEE counted 1,500 of the agency's nearly 40,000 employees as members. In early 1990, DeBonis quit the agency to concentrate full-time on organizing.

Forest Service employees say AFSEEE's activities are having an effect. The Forest Service for a long time has been "very commodity-oriented," says Gary Cordova. "Now, we're slowly coming around to having a balanced program."

TIMBER CUTTING'S ENVIRONMENTAL LEGACY

The balanced program that Cordova is referring to is supposed to include recreation, wildlife and fish preservation, and management of soil and water resources. All too often, careless cutting—on both private and public lands—has left a legacy not only of shorn hillsides, but of mudslides, erosion, ruined streams, and damaged fisheries. For example, Forest Service researchers have found that several river systems in the Columbia River Basin, where there has been extensive logging, farming, and grazing over the past few decades, have lost 50% to 75% of their large pools over the past half century. These pools are vital for salmon because they serve as resting spots for adults before they spawn, as rearing areas for young fish, and as refuges during droughts and winter freezes.

Although attention has focused on the old growth west of the Cascades, the damage extends eastward across Washington and Oregon into the forests of Idaho and Montana as well. In the past few years, the Plum Creek Timber Company has rapidly leveled forests on much of the 1.4 million acres it owns in the Northwest. The results often have not been

NEW FORESTRY

When is a clear-cut not a clear-cut? When it has a few trees left on it. Welcome to "new forestry."

The traditional method of logging is to level every tree on a particular site, and then clean up the remaining debris to prepare the ground for replanting. The forest industry argues that clear-cutting is good forestry because it mimics the results of "natural occurrences" such as fires, and maximizes timber yields. But there are several arguments against this method, not the least of them the ugly, gaping holes that clear-cuts leave. Critics contend that clear-cuts exacerbate erosion, ruin soil quality, and destroy wildlife habitat. Ultimately, the diversity of natural forests is lost, because land cut this way is generally replanted with single-species forests of uniform age.

An alternative method is to leave some live trees, dead standing trees (snags), and fallen logs. This is not much prettier than a clear-cut, but proponents argue that it has the possibility of regenerating into more diverse second-growth forest that can provide habitat to old-growth denizens such as the northern spotted owl. The University of Washington's Jerry Franklin, the father of new forestry, describes it as an attempt to "manage ecosystems for their multiple values rather than to focus on individual resources, whether wood or owls."

New forestry has its detractors on both sides of the aisle. Some environmentalists deride it as putting a little makeup over ugly and destructive clear-cuts. The Wilderness Society's Jean Durning emphasizes that new-forestry techniques are only an alternative way of managing commercial

> timber, not a replacement for old-growth forests. Many forest industry folks, too, are skeptical of its widespread application. "There's nothing wrong with Jerry Franklin's philosophy," says Robert Gustavson, director of forest management for the Washington Forest Protection Association. "There are a lot of places it would be useful, like providing more old-growth habitat." But Gustavson takes a dim view of its commercial viability: "If you had to manage private lands that way, it would take away all your incentives."
>
> Still, some firms, notably the Plum Creek Timber Company, are experimenting with new forestry. The Forest Service is actively pushing it, and the Washington State Department of Natural Resources plans to experiment with alternative techniques in forest management on a large experimental forest on the Olympic Peninsula.

pleasant. In 1988, a Plum Creek clear-cut extensively damaged a tributary of the Swan River in northwestern Montana. Erosion from road building and logging buried trout spawning grounds under a layer of sediment. State scientists studying the damage found that westslope cutthroat trout eggs would suffer a 96% mortality rate. The damage is expected to be even greater for bull trout, a species "of special concern" in Montana. "Based on predicted embryo survival, bull trout . . . will experience 100% mortality in sampled spawning areas" on Jim Creek, states a report by researchers from the Montana Department of Fish, Wildlife and Parks. The importance of Jim Creek is underscored by the fact that it is one of only about a dozen streams in the Flathead Basin that serve as the primary spawning grounds for bull trout.

The timber industry maintains that incidents like the one at Jim Creek are the exception, and that its forest practices have become much more environmentally conscious in recent years in response to tighter regulations, more public scrutiny, and increased knowledge of the effects of logging on ecological systems. "What you see today is the result of a lot of past practices," says Robert Gustavson, director of forest management for the Washington Forest Protection Association (WFPA), a timber-owners' organization whose members control about half the private timberland in the state. "I can remember 20 years ago when we used to put bulldozers in the creeks [to clear out fallen logs and other debris], because that's what the fisheries people told us to do. What we do changes with knowledge." Now, for instance, loggers are supposed to leave uncut buffer zones along streams to protect spawning grounds and water quality.

Certainly, there have been some changes on the part of the industry. One notable example is provided by Plum Creek, whose penchant for clear-cutting has earned it a terrible reputation. ("Within the industry, they're considered the Darth Vader of the state of Washington," was the comment of one Washington congressman.) Plum Creek is now one of the leading experimenters in "new forestry" techniques, which are designed to provide an alternative to clear-cuts. According to Gustavson, industry foresters are also looking into an approach to timber management called "ecological landscaping," in which timber-management efforts take into account the interaction of forest practices and the ecology of large areas such as river drainages. Still, as much as the forest industry has begun to change its techniques, the debate over whether private-industry forest practices are providing enough protection for wildlife, fisheries, and water resources continues unabated.

AN ATTEMPT AT NEGOTIATION FALLS APART

In Washington State, environmentalists, Native American tribes, and the timber industry, all worn down by years of acrimonious disputes, agreed in the mid-1980s to what one newspaper termed "a historic new deal for Washington forests." It was a process called Timber, Fish, and Wildlife (TFW). TFW brought representatives of environmental groups, timber industry officials, Indian tribal leaders, and state officials from agencies such as the Department of Wildlife and the Department of Natural Resources (DNR) to the negotiating table to review management of state and private timberlands. Among TFW's most significant accomplishments were the institution of an annual review of timber companies' harvest plans to assess environmental impacts, and a procedure for site review of applications for timber-cutting permits. The process was described in an article in *The Northwest Environmental Journal* (Spring/Summer 1990): "These reviews, by so-called interdisciplinary teams, bring together representatives of TFW cooperators at the site. There, they negotiate with the landowner and one another to determine the conditions to be applied to the permit by DNR. It is at this point that tree-by-tree decisions can be made to conserve habitat or Indian cultural values, while proceeding with timber harvest." To implement and enforce TFW agreements, state natural resource agencies beefed up their staffs, and environmental organizations and Indian tribes received funding to participate in the process.

For a while, the negotiations seemed to be proceeding fairly well. But then came an incident that served to reveal Timber, Fish, and Wildlife's barely concealed problems. In 1989, two landowners applied to the Washington Department of Natural Resources for permits to log large tracts of timber near Lake Roesiger in Snohomish County. Nearby residents were appalled at the prospect of such a large cut, and environmentalists were concerned that the rapid harvest of timber would adversely affect wildlife, fisheries, water quality, and soil stability in the area. They also worried that logging might be a prelude to commercial development of the land. Snohomish County entered the fray, going to court to seek a more comprehensive review by the DNR of the environmental effects of the logging.

> *If one does not realize how sinful man's behavior has been toward nature's abundance, let him visit what were once the proud forests of the West.*
>
> —ANGELO PELLEGRINI
> THE UNPREJUDICED PALATE, 1948

The Lake Roesiger incident revealed ongoing problems that TFW had not yet dealt with, including rate of harvest, wildlife habitat protection, and conversion of timberland to other uses. Timber, Fish, and Wildlife had put off addressing these issues because they were the most contentious. But the Lake Roesiger case, and other incidents in which logging operations near residential areas had raised the hackles of the locals, forced the hand of the TFW negotiators. A booming timber market in the late 1980s had accelerated the pace of cutting, and a hot real estate market made it increasingly attractive to turn forests into subdivisions.

In late 1989, state commissioner of public lands

Brian Boyle, who heads the DNR, helped convene the Sustainable Forestry Roundtable (SFR) to try to resolve these problems. The feeling was that a forum separate from TFW was needed to focus on these tough questions, even though the participants in the Roundtable were largely the same as in Timber, Fish, and Wildlife. After months of discussions, the negotiators thrashed out a proposal that called for restraints on the size of clear-cuts, and for the timber industry to set aside 10% of its lands—about a million acres—for wildlife. The agreement, which was to last 10 years, also mandated government reviews in cases in which cutting covered more than 4% of a watershed, and prescribed financial penalties to discourage the conversion of forestland to non-timber uses.

But the proposed agreement never got off the table. Although the forest industry, led by the Washington Forest Protection Association, overcame some internal opposition and approved the proposal, the environmentalists were badly split. The Washington Environmental Council, which represents dozens of organizations, accepted the plan, but only with many modifications. The National Audubon Society and a Kittitas County group called Ridge rejected it, while environmentalists from Whidbey Island voted to abstain. The environmentalists generally maintained that the proposal didn't go far enough to protect wildlife habitat and streams.

The environmental community was immediately lambasted in the press for torpedoing the agreement. The timber industry emerged with a much-improved public image; it was portrayed as willing to compromise, while the environmentalists were seen as having arrogantly rejected a measure that would have revamped forestry practices in Washington.

Timber 61

The truth is probably not so black-and-white. In fact, environmentalists had given up quite a lot in the negotiations. The timber industry had succeeded in getting them to shift quite a bit from their original positions, and had ended up with much of what it wanted in the final proposal. "Things kept appearing [in the proposal] that we never agreed to," asserts Jim Pissot, director of the National Audubon Society's Washington office. "We got bamboozled and we were stupid." Many environmentalists felt, for example, that a greater percentage of forestland should have been put aside to preserve wildlife, and that the rate of harvest allowed by the proposal was still too rapid. Pissot acknowledges that the environmentalists made a mistake in not more actively promoting their side of the story. "Right after the proposal was rejected, I should have started calling newspapers around the state. But I didn't," he says. "Frankly, it was a matter of time, and I'm dealing with a half dozen different issues."

Whatever the true story, the whole process quickly disintegrated. "It's over. It's done with. They killed it," said William Jacobs, executive director of the Washington Forest Protection Association, after environmental organizations rejected the proposal. Brian Boyle and Governor Booth Gardner tried to pick up the pieces by introducing forest-reform legislation based largely on the rejected plan. But both sides attacked it. Environmentalists were critical because it weakened wildlife-protection provisions. The WFPA opposed taxing wood and paper products to fund it. The industry apparently felt confident enough of the considerable power it wields in the state legislature to abandon Boyle's bill and push its own version, which did little to protect wildlife or provide for harvest monitoring or funding. A third proposal, which would have reorganized the state board that regulates forest practices, was deemed too radical by the timber industry.

A stalemate quickly developed, and the legislature currently seems as unable to deal with the tough questions of forest management as the Sustainable Forestry Roundtable process was before it. After almost two years of SFR and legislative fighting, all sides are looking once again at Timber, Fish, and Wildlife as a means toward a workable agreement. But time isn't the only thing that has been lost; trust has been damaged among the TFW participants. As for the Lake Roesiger tracts, substantial areas have already been cut.

WHITHER TFW?

It is a sunny spring morning, but the mood inside the small room at the Federal Way (Washington) Public Library is decidedly gloomy. About 20 participants in the Timber, Fish, and Wildlife negotiations are glumly munching muffins, downing coffee, and discussing the future of the process. The meeting is taking place at a particularly dark time for TFW: the Sustainable Forestry Roundtable negotiations have recently fallen apart, and it is pretty clear that the legislature is equally unable to move things forward. Now questions are being raised about the future of the TFW process itself.

The group has to contend with a feeling among some observers that TFW, despite producing some reforms, is long on process but short on results. As one participant puts it, Timber, Fish, and Wildlife is seen as "a great place to flap your jaw, but nothing comes out of it." Jim Pissot is clearly impatient with the situation. "I want to get from third base to

> **RANGE WOES**
>
> Forests are not the only places where federal land management is coming under fire. Environmentalists charge that ranchers are allowing their livestock to overgraze rangeland owned by the Forest Service and the Bureau of Land Management, thus damaging wildlife, vegetation, soil, and water. The amount of land involved is far from insignificant; the Forest Service and the BLM control more than 300 million acres in the West, and almost 90 million acres in the northwestern states.
>
> Critics of the federal grazing program also contend that Uncle Sam subsidizes destructive overgrazing by charging extremely low rates for the lease of government land. A report released in 1991 by the congressional General Accounting Office points out that grazing fees cover only a small fraction of the taxpayer cost of running the program, and legislation has been introduced in Congress to dramatically increase those fees. Ranchers in the Northwest would surely feel the pinch of such increases, particularly in Montana and Idaho, where nearly 30% of the nearly 32,000 ranchers lease federal grazing land.
>
> Ranchers contend that if grazing fees are pumped up, many of them will be forced out of business. Some aren't slow to compare ranchers to timber workers: "The grazing fee issue is just a means to an end—and the end is get us off the public lands just like the spotted owl is getting timber production halted," one livestock industry representative told *USA Today*.

home," he tells the gathering in frustration.

The Sustainable Forestry Roundtable's failure

has exacerbated feelings of disappointment with the negotiation process and damaged much of the trust built up previously through TFW discussions. The Roundtable collapse was a "setback for TFW," Bill Jacobs acknowledges. "So much attention was taken up by SFR that TFW was neglected. We have a lot of catching up to do."

Criticism of the Timber, Fish, and Wildlife process doesn't stop at its failure to deal with such difficult questions as wildlife habitat preservation and rate of harvest. Some TFW participants are also critical of a failure to monitor compliance with forest-practices regulations. "A key component to the success of TFW is the recognition that the TFW process is effecting change in the field," an October 1990 TFW Field Implementation Committee report notes. "However . . . it is perceived by many that routine compliance monitoring is, at best, minimal." The report states that 125 forest-practice applications were supposed to be reviewed in 1990, but that only 49 were.

In his Olympia office, Jim Pissot picks up a copy of the implementation report and begins flipping through the pages. "My job is to get results," he says. "That means evidence that the regulations are being implemented, that monitoring is occurring, and that there is compliance." Pointing out several gaps in the data, he adds, heatedly, "I don't see that here."

A former state official made similar charges a couple of years ago. In a 1989 talk in Seattle to wildlife and fisheries managers, Eugene Dziedzic, former Chief of Habitat Management for the Washington Department of Wildlife, commented that "corporate timber can do just about anything it wants to do by following the few standards they accepted. Cutting on state and private lands is near or at record levels

. . . . But, to date, no one has really checked to see how much has been gained by wildlife and other interests involved in TFW."

Industry representatives, while claiming that compliance is widespread, acknowledge the lack of monitoring. Jacobs calls DNR's oversight "spotty," citing inadequate funding for state agencies. The industry appears to want the Timber, Fish, and Wildlife process to continue. Jacobs thinks that TFW can accomplish a lot of what the Sustainable Forestry Roundtable failed to do, if it doesn't attempt to address too many issues at once—something he sees as a major failing of SFR. He acknowledges the impatience of a number of participants, particularly environmentalists. But he contends that many of the critics "are new to the process" and have not recognized many of the successes that TFW has achieved.

With SFR dead and the state legislature paralyzed for the time being, the pressure on Timber, Fish, and Wildlife is growing. If it doesn't produce, there's a good chance that environmentalists will throw down the gauntlet again. Pissot, for one, makes it clear that he wouldn't hesitate to come to legal fisticuffs with the timber industry if TFW were to stumble badly. The Washington Environmental Council's Marcy Golde noted how shaky the timber ceasefire is: "It's indicative of how deep a hole we are in that there is a willingness to consider that conflict is more productive than this process."

TIMBER'S FUTURE

The timber industry has always been cyclical. The post–World War I housing boom helped spur an upsurge in the Northwest wood products market, and the amount of lumber produced in Washington reached an all-time high of 7.5 billion board feet in 1926. Then, in just a few years, that figure plunged to 2.2 billion board feet as the industry felt the icy grip of the Great Depression. The aftermath of World War II ushered in another strong market, and harvests in national forests grew rapidly until the 1960s. The recession of the early 1980s hit the timber industry hard; wood products employment in Washington dropped about 30%. But the timber market picked up dramatically in the latter half of the decade, fueled in large part by foreign demand. Demand has once again weakened in the recession of 1990–1991.

There is no doubt that the industry's volatility will continue in the 1990s. But, boom or bust, it can count on being squeezed by a dwindling supply of timber, particularly from federal land. Oregon will no doubt suffer most, as more than half its total harvest comes from federal lands, compared with less than 25% in Washington. To make matters worse, private and state-owned lands aren't about to make up the deficit over the next two decades; the timber industry, for the most part, didn't start widespread replanting until the 1960s. (There are exceptions: industry giant Weyerhaeuser started reforesting much earlier.) And because of the 50-year growing cycle, most of those trees won't be ready for harvest for a long time. Harvests from private timber-industry lands in western Washington and Oregon are expected to decline by 8% to 14% over the next two decades, according to one estimate. In western Oregon, where harvests dropped

> *What is man without the beasts? If all the beasts were gone man would die from a great loneliness of spirit. For whatever happens to the beasts soon happens to man. All things are connected.*
>
> —CHIEF SEATTLE, 1854

by 13% during the mid-1980s over those of the previous decade, the rate of decline "is expected to accelerate sometime during the next decade," according to the Oregon Department of Forestry. The total harvest "will continue to decrease through 2015."

These somewhat abstract figures translate into pain for timber-dependent communities. Three Seattle economists (Glenn Pascall, Dick Conway, and Douglas Pedersen) predict that Washington's wood products industry will shed almost 10,000 jobs during the 1990s—a plunge of more than 20%. Most of that will occur in the first half of the decade, when the effects of a slowing economy and the reduced supply of federal timber—the study assumes a 50% cut—will be felt.

But environmentalists and growing numbers of economists contend that the timber industry's longer-term woes are largely unrelated to the supply of federal timber. Spotted owl protection may result in a more rapid decline in timber jobs, but it is likely that most of those jobs would have disappeared sooner or later as the remaining old-growth timber was logged. Growing productivity will also take its toll. As Judge Dwyer observed, "Job losses in the wood products industry will continue regardless of whether the northern spotted owl is protected. A credible estimate is that over the next twenty years more than 30,000 jobs will be lost to worker-productivity increases alone." Unfortunately, this argument does little to alleviate the pain or ease bitter feelings in timber-dependent areas such as the Olympic Peninsula or Douglas County, Oregon. With supply being squeezed as it is—and likely to become more constrained—the prospects for peace in the woods seem to have dimmed, as environmentalists fight to protect forests from rapid cutting, and industry tries to keep its grip on as much timber as possible.

A shortage of logs is not the only pressure the industry faces. Competition from foreign producers and from other parts of the country—particularly the South, where the growing cycle is much faster—is likely to increase in the coming years. More stringent environmental protection will also take a bite out of the industry. All of these factors are likely to force smaller, less-efficient producers out. The giants of the industry, like Weyerhaeuser, which owns about 1.5 million acres of timberland in Washington alone, are well positioned in this uncertain economic and political climate. In fact, limits on federal timber will in all likelihood be extremely beneficial to large companies that own extensive tracts of timber, because they will tend to drive prices up.

Another pressure is building on both timber supply and wildlife habitat: the continuing disappearance of forestlands. Between 1930 and 1980, 4 million acres of Washington forests were converted to non-timber uses. During the 1980s, about 80,000 acres were converted. That trend seems destined to continue. As growth has propelled development farther and farther from metropolitan centers like Seattle and Tacoma, the price of land in formerly rural areas has shot up. Many timber owners are now in a position to make a quick killing selling their property to developers rather than waiting 50 years for the next timber harvest. Washington commissioner of public lands Boyle predicts that the central Puget Sound area will lose another 80,000 acres of forestland over the next 10 years.

The timber industry in the Pacific Northwest is not about to disappear, but it is certain to change. It is in the midst of a transition that has been forced on it by shrinking supply and overcutting. Citizen concern and activism have already played a major role in preserving ancient forests and changing the way in which the timber industry operates. In the coming years, it's likely that the timber industry and natural resource agencies such as the Forest Service will change even more. "A shift in agenda is needed. Industrial users must recognize that society views forestland as more than agricultural land with a slow-maturing crop, and it expects more of that land," Jerry Franklin writes. "Conversely, environmentalists must move away from preservation as the sole solution for many social objectives.... Hence, management of this commodity land is critically important to us all." Residents of the Northwest should be active in that shifting agenda. It's the least we owe to a resource that not only has helped to shape the region's economy and history, but has given it much of its identity.

RESOURCES

ALLIANCE FOR THE WILD ROCKIES
Box 8731
Missoula, MT 59807
(406) 721-5420

See listing in Growth chapter.

ALPINE LAKES PROTECTION SOCIETY
Route 1, Box 890
Ellensburg, WA 98926
(206) 774-5047

Works with the U.S. Forest Service on the management of a sizable portion of wilderness acreage near Snoqualmie Pass.

ASSOCIATION OF FOREST SERVICE EMPLOYEES FOR ENVIRONMENTAL ETHICS
PO Box 11615
Eugene, OR 97440
(503) 484-2692

See description in this chapter.

CASCADE HOLISTIC ECONOMIC CONSULTANTS (CHEC)
14417 SE Laurie Avenue
Oak Grove, OR 97267
(503) 652-7049

Nonprofit forestry consultant offering educational and technical assistance to conservation groups and state agencies. Publishes *Forest Watch*, a magazine covering forest planning and management.

GREATER YELLOWSTONE COALITION
PO Box 1874
Bozeman, MT 59771
(406) 586-1593

Focuses primarily on public land use, with limited work on private lands. The coalition targets threatened and endangered species, oil and gas development, mining, timber, and river protection in the greater Yellowstone ecosystem. A small staff and many volunteers working on specific projects

Timber Resources 71

(writing letters, testifying at hearings) help organize communities and smaller environmental groups around important issues.

IDAHO CONSERVATION LEAGUE
PO Box 844
Boise, ID 83701
(208)345-6933

See listing in Water chapter.

IDAHO WILDLIFE COUNCIL
PO Box 7043
Boise, ID 83708
(208)344-5159

See listing in Growth chapter.

INLAND EMPIRE LANDS COUNCIL
PO Box 2174
Spokane, WA 99210
(509)327-1699

A community organizing group that concerns itself with all forestry issues, including water quality, hunting, fishing, mining, and road construction. Gets the public involved with the implementation of individual forest plans, appealing unacceptable timber sales and making sure the Forest Service holds to its multiple-use mandate. Focuses on eastern Washington and northern Idaho, but would like to expand into northwestern Montana, eastern Oregon, and more of Idaho. Two staff members supplemented by volunteers, in the office and in the field.

LIGHTHAWK: "THE WINGS OF CONSERVATION"
340 15th Avenue E
Seattle, WA 98112
(206)324-5338

A group of staff and volunteer pilots who fly politicians, educators, media, activists, and other environmental organizations over national forests for a firsthand look at the effects of clear-cutting.

Montana Wilderness Association
PO Box 635
Helena, MT 59624
(406) 443-7350

The oldest conservation group in the state (30 years) works mainly on lands unprotected by wilderness designation, some BLM lands, and rivers. The association is backing the the Kootenay-Lolo Accords (landmark agreements between timber and mill workers and conservationists on wilderness preservation). Extensive education program. A strong membership base and some volunteers.

Montana Wildlands Coalition
PO Box 213
Helena, MT 59624
(406) 442-2566

An association of 45 local and state organizations dedicated to wilderness designation for the remaining roadless lands in Montana. Efforts are aimed at organizing local people, as well as persuading state representatives to introduce legislation. Is very involved in the negotiation and implementation of the Kootenay-Lolo Accords.

The Mountaineers
300 Third Avenue W
Seattle, WA 98119-4100
(206) 281-8509

A membership organization concerned mainly with ancient forest protection, but also involved in wetlands, rivers, fish, and wildlife. Its conservation orientation is for the most part recreational, and a great deal of effort goes into educational field trips, hikes, etc. Monthly magazine *The Mountaineer* has an activities list and environmental issue coverage.

National Audubon Society

Washington State Office
PO Box 462
Olympia, WA 98507
(206) 786-8020

Western Region Office
555 Audubon Place
Sacramento, CA 95825
(916) 481-5332

Rocky Mountain Regional Office
4150 Darley Avenue, Suite 5
Boulder, CO 80303
(303) 499-0219

Northern Rockies Natural Resources Center
240 N Higgins Street
Missoula, MT 59802
(406) 721-6705

Pacific Northwest Natural Resources Center
519 SW Third Avenue, Suite 606
Portland, OR 97204
(503) 222-1429

See listing in Growth chapter.

NATIVE FOREST COUNCIL
PO Box 2171
Eugene, OR 97402
(503) 688-2600

Supports a no-compromise forest protection bill, NFPA 91, which it's pushing through letter writing, lobbying, and a national media campaign. Also strong in education, sending newsletter *The Forest Voice* to schools and libraries.

OREGON NATURAL RESOURCES COUNCIL
522 SW Fifth Avenue, Suite 1050
Portland, OR 97204
(503) 223-9001

See listing in Water chapter.

SELKIRK PRIEST BASIN ASSOCIATION
PO Box 181
Coolin, ID 83821-0181
(208) 448-1813

See listing in Water chapter.

Sierra Club

Cascade Chapter
1516 Melrose Street
Seattle, WA 98122
(206) 621-1696

Montana Chapter
c/o James Conner
78 Konley Drive
Kalispell, MT 59901
(406) 752-8925 or contact the field office in Sheridan, WY at (307) 672-0425.

Northern Rockies Chapter
c/o Edwina Allen
1408 Joyce Street
Boise, ID 83706
(208) 344-4565

Oregon Chapter
c/o John Albrecht
3550 Willamette Street
Eugene, OR 97405
(503) 343-5902

See listing in Growth chapter.

Sierra Club Legal Defense Fund

Northwest Office
216 First Avenue S, Suite 330
Seattle, WA 98104
(206) 343-7340

Rocky Mountain Office
1631 Glenarm Place, Suite 300
Denver, CO 80202
(303) 623-9466

See listing in Growth chapter.

Washington Environmental Council
4516 University Way NE
Seattle, WA 98105
(206)527-1599

See listing in Growth chapter.

Washington Forest Protection Association
711 Capitol Way, Suite 608
Olympia, WA 98501
(206)352-1500

An association of private commercial forest landowners who own about half of the timberland in Washington State. Active in the Timber, Fish, and Wildlife negotiations. Also conducts tours of timber facilities and outreach activities. Emphasis is on getting industry's side of the timber controversy out to the public.

Washington Wilderness Coalition
PO Box 45187
Seattle, WA 98145-0187
(206)633-1992

See listing in Growth chapter.

The Wilderness Society

Idaho Chapter
413 W Idaho Street, Suite 102
Boise, ID 83702
(208)343-8153

Montana Chapter
105 W Main Street, Suite E
Bozeman, MT 59715
(406)586-1600

Oregon Chapter
610 SW Alder Street, Suite 915
Portland, OR 97205
(503)248-0452

Washington Chapter
1424 Fourth Avenue
Seattle, WA 98101
(206) 624-6430

Focus is on the ancient forest. Most attention is paid to federal lands, although each regional office also does some work on a state level. Washington and Oregon have already passed comprehensive wilderness designation bills (in 1984), but environmental groups in Montana and Idaho are still campaigning and negotiating. Has been instrumental in mapping out what remains of the region's old growth. Monitors timber sales and forest plans and appeals those it considers environmentally damaging. The society's land-protection ethic extends to offshore drilling, continental shelf issues, and the BLM's management of wildlands. Activities run from contact with communities through media to direct lobbying of Congress. The Idaho office is also involved with water quality issues and petitioning for endangered species listing for wild salmon.

WATER

THE BATTLE OVER WATER

THE ENVIRONMENTAL BILL

DECADES OF FAILED PROTECTION

WESTERN WASHINGTON WATER WOES

WATER QUALITY

FROM GROUND
TO TAP: YOUR DRINKING WATER

SURFACE WATER

NONPOINT POLLUTION

CRACKS IN WATER QUALITY ENFORCEMENT

WHAT YOU CAN DO

WATER

Many a winter's day leaves denizens of the sodden western side of the Cascade mountains wondering if the rain and clouds will ever go away. With precipitation at least 150 days of the year in that area, it does not seem that there could be a shortage of water in the Northwest. Plentiful irrigation water has turned the region's deserts into an agricultural breadbasket. Cheap hydroelectric power has kept the Northwest's electricity rates far below those of the rest of the country.

We have so much water that parched southern California covets some of our bounty. When the Los Angeles Board of Supervisors made an ineffectual grab for water from the Columbia and Snake rivers in 1990, it was the object of derision from the Northwest. But now we are facing the fact that it isn't Californians who pose the real threat to our rivers, it is northwesterners. American Rivers, a conservation group based in Washington, D.C., lists the Columbia and Snake among the 10 most endangered rivers in the United States. For too many years, we ignored the warning signs. Now the region's inhabitants are beginning to pay for that oversight.

THE BATTLE OVER WATER

The Columbia River is one of the great natural landmarks of the Northwest. Its journey of more than 1,200 miles to the Pacific starts in British Columbia. From there it flows south into Washington State until it meets the Spokane River and turns west. Before it reaches the Cascade Range it turns south again, passing the Hanford nuclear facility and joining with its main tributary, the Snake, flowing from Idaho.

Soon after that, it turns west once more to form the border between Washington and Oregon all the way to the Pacific.

Today, the Columbia no longer flows freely to the ocean; it has been turned into an immense power plant and a giant watering can. The region's dependency on the Columbia and its tributaries is clear from the numbers: about a dozen dams cut across its path, providing electricity to homes and businesses and irrigation water to arid lands. In all, there are 136 dams in the Columbia River Basin. The Columbia River system provides about two-thirds of the electricity used in the Northwest and waters millions of acres of farmland.

There are of course legends about the salmon and these falls. The Bridge of the Gods, which long ago spanned the Columbia shortly above Bonneville, fell into the river, creating a dam which the salmon could not negotiate. Coyote saw the peril to the people who depended so much on salmon for food. He went to work, clearing out a channel through the barricade so the salmon could once more seek the headwaters. Then to complete the job he went down the river and herded the salmon upstream through the channel.

—WILLIAM O. DOUGLAS
OF MEN AND MOUNTAINS, 1950

The river has paid a price for such generosity. The great dams that were built beginning in the 1930s and 1940s sacrificed the salmon. The construction of the Grand Coulee Dam alone wiped out more than a thousand miles of salmon spawning grounds. Today, the Columbia and the Snake rivers are so modified that they are inhospitable environments for the fish. During the spring, the water in the Snake takes two weeks to reach the Pacific from Lewiston, Idaho. When white settlers first arrived in the region, it took only two days.

Each year, salmon make their way from spawning grounds hundreds of miles upriver to the Pacific Ocean, a journey whose timing is intricately tied to their biology. Juvenile salmon, genetically adapted to the temperature and velocity of an unmolested river, are slowed by warm, sluggish water behind massive dams, or get chewed up by power turbines. The slow-moving water also exposes them to diseases and predators for longer periods of time. The result is that fewer and fewer salmon survive to return later on the epic trip upstream to their spawning grounds. Only 2.5 million salmon are left in the Columbia—an 85% drop since the late 1800s—and only a fraction of those are wild fish; the rest come from hatcheries. It is estimated that hydropower development and operation account for 80% of that decline.

The damage to wild fish runs extends over much of the Northwest. More than a hundred "major populations" of salmon and steelhead (a large anadromous trout) on the West Coast have already been wiped out, according to a 1991 American Fisheries Society survey. Some 101 species of naturally spawning salmon, steelhead, and sea-run cutthroat trout stocks native to the Northwest and California are at high risk of extinction, and another 58 species are at moderate risk.

Lower Granite Dam Sockeye Counts

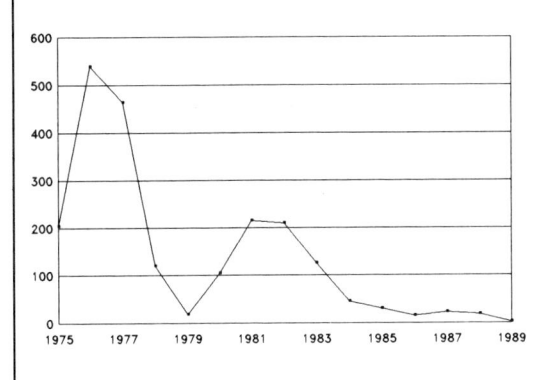

National Marine Fisheries Commission

The dams are not the only factor in the decline of the salmon; biologists also blame agricultural and logging practices, and overfishing by commercial and sports fishermen. "Everybody thinks it's just dams. It ain't," says Andy Kerr, of the Oregon Natural Resources Council. "It's also habitat destruction, overharvesting, and competition from hatchery fish." (Hatchery fish have depleted the gene pools of the wild salmon and carry diseases to which wild fish are susceptible.) With regard to the river, "Everybody has been maximizing their self-interests," Bill Bakke, executive director of the conservation group Oregon Trout, told *The Oregonian* in 1990. "Now the environmental bill is coming due."

THE ENVIRONMENTAL BILL

That environmental bill is being called in by the Snake River sockeye salmon. The obstacles the Snake River sockeye faces every spring on the nearly 1,000-mile journey from its spawning grounds in Idaho down the Salmon, Snake, and Columbia rivers have finally become practically insurmountable. Some 531 sockeyes were sighted in 1976 at the Lower Granite Dam, about 30 miles from the Washington-Idaho border. By 1989, that number had declined to two. The following year, not a single sockeye was sighted either at the Lower Granite Dam or at Redfish Lake, the site of the fish's sole remaining spawning ground. Ironically, Redfish Lake took its name from the color of the spawning salmon.

In March 1990, the Shoshone-Bannock Tribes of the Fort Hall Reservation in Idaho petitioned the National Marine Fisheries Service (NMFS), an arm of the U.S. Department of Commerce, to list the Snake River sockeye as an endangered species.

MILITARY WASTE

While the military was arming itself for the Cold War, it was also becoming the nation's largest polluter. Currently, the U.S. Department of Defense (DOD) is in the process of investigating more than 17,000 suspected toxic waste sites at 1,855 installations such as air bases, naval stations, and munitions plants. In addition, the Pentagon is trying to assess the extent of contamination at about 7,000 facilities it formerly owned or used.

This is the legacy of years of carelessly handling huge amounts of waste produced in operations that vary from airplane cleaning to the manufacture of munitions and nerve gas. These activities generate huge amounts of waste—nearly a billion pounds in 1989 alone. The contaminants found at DOD installations range from seemingly ubiquitous solvents like trichloroethylene and heavy metals to more exotic compounds such as TNT and diisomethylphosphonate (DIMP), a by-product of nerve gas production. In state after state, chemicals from military bases have found their way into soil, air, and water. In addition, unexploded ordnance is littered across large tracts of land. About a hundred DOD sites have been placed or proposed for inclusion on the EPA's high-priority Superfund hazardous waste cleanup list. Nine of these DOD Superfund sites are located in Washington State. One northwestern site that causes special concern is the Umatilla Army Depot, in northern Oregon, where several hundred thousand gallons of nerve gas and chemical warfare agents are stored.

Department of Energy (DOE) nuclear weapons facilities are also environmental disasters, at which the problem of chemical contamination is often compounded by radioactive waste. The DOE's sites, though fewer than the Pentagon's, tend to be much larger, and the amount of waste sometimes mind-boggling. At the 560-square-mile Hanford Nuclear Reservation, in southeastern Washington, radioactive and chemical waste from nearly 50 years of operation has contaminated approximately 1,400 sites. According to the DOE's own estimates, there's enough radioactive waste alone at Hanford to cover a football field to a depth of 700 feet. Billions of gallons of liquid waste have been dumped, and groundwater pollution has spread over about 200 square miles. Contamination has been detected in the nearby Columbia River, though the DOE maintains that the levels are not hazardous and are below legal limits.

The bill for repairing the environmental damage at federal facilities is going to be astronomical. One estimate puts the total cost, including that for nonmilitary facilities, at $330 billion. Hanford's cleanup tab alone is estimated to reach $57 billion over the next three decades.

But enforcement of cleanup efforts at federal facilities can be problematic. Take the Hanford facility. In 1989, the state of Washington, the EPA, and the DOE reached an agreement that laid out a timetable for accomplishing the massive task, a schedule that was supposed to be legally enforceable. But in early 1991, the DOE announced it wanted to delay some phases of the cleanup. Both the EPA and the state of Washington reacted with outrage. "It's astonishing that Energy would unilaterally let such a major milestone slip," declared EPA regional administrator Dana Rasmussen. "I hope your action does not signal a return to the days of noncommunication and noncooperation," echoed Washington governor Booth Gardner in a letter to DOE secretary James Watkins.

That cooperation is necessary because environmental agencies have a limited ability to fine or bring action against federal facilities for violations of environmental laws. The U.S. Department of Justice, for instance, will not bring suit against another federal agency. According to a June 1990 survey by Friends of the Earth, DOD facilities in Washington would have been subject to tens of thousands of dollars in penalties for environmental violations in the past couple of years alone, if they had belonged to a private company. That leaves regulators in the position of relying on negotiation to bring the military into compliance with environmental laws. Says Dyan Oldenburg, formerly of the Military Toxics Network, "It's unconscionable that there should be this double standard."

"People are starting to take this more seriously," Gordon Davidson, the acting head of the EPA office responsible for monitoring federal facilities, told *Newsday*. But "we've still got a long way to go." Environmentalists are largely unimpressed. Says Oldenburg, cleanup efforts are "moving so slowly, it's pitiful."

One year later, in April 1991, the NMFS announcement that it would propose the Snake River sockeye for listing as an endangered species didn't really surprise anyone. The region seemed to be getting used to endangered species. FIRST THE OWL, NOW SNAKE RIVER SALMON, blared the *Seattle Times* headline. But if the NMFS carries through on its proposed listing, the recovery plan could affect many more people than those touched by efforts to protect the beleaguered spotted owl.

One measure under consideration is the elimination of the sockeye catch by both tribal and nontribal fisheries in the Columbia. Other possible steps include controlling predator populations and placing screens over turbine intakes and agricultural diversions to prevent fish from getting trapped in them. Although the NMFS and the states have been working at this screening for several years, the job is far from done.

Any serious attempt to save the Snake River sockeye will profoundly transform the way the water in the Columbia River system is managed. That is going to mean big changes for water users, including farmers, electrical utilities (and thus ratepayers), fishermen, and barge operators. There is simply not enough water to generate all the electricity the Northwest wants and to irrigate millions of acres, while assuring the survival of enough salmon to propagate the species.

The Columbia Basin Fish and Wildlife Authority, which represents regional fish and wildlife agencies and Native American tribes, has asked for large increases in the amount of water released on the Snake and Columbia during spring and summer months, to assist salmon and steelhead migration. That, however, is the time of year when power officials want to store as much water as possible, to make it available during the high-demand winter months. Power industry interests claim that draining volume from storage reservoirs could result in the loss of several thousand megawatts of power. (Seattle runs at about a thousand megawatts a day.)

This is not an academic debate. The vaunted power surplus that existed in the Northwest during the 1980s has disappeared. Any lost hydroelectricity would have to be made up from other sources, potentially at a much greater cost to the public, which in turn could mean rate shock for Northwest utility customers and industrial users who consume huge amounts of cheap hydropower. Particularly vulnerable to rate increases is the region's aluminum industry, which employs about 10,000 people and produces nearly half of the nation's aluminum.

Farmers will also pay a price to save the salmon. Agriculture accounts for about 75% of the water used in Washington and 90% in Oregon, and most of that comes from rivers in the Columbia Basin. Water diverted for agricultural use from the Snake River amounts to about a sixth of the river's annual flow—a total of 6 million acre-feet (the amount of water it takes to cover 6 million acres with one foot of water). Agricultural use is partly responsible for draining some streams and rivers dry at certain times of the year.

> Agriculture accounts for about 75% of the water used in Washington and 90% in Oregon.

DECADES OF FAILED PROTECTION

The recognition that dams threaten salmon stocks is not new. A report presented at an American Fisheries Society meeting in 1937 noted, "That part of the industry dependent on the Columbia River salmon run has expressed alarm at the possibility of disastrous effects upon the fish through the erection of the tremendous dams at Bonneville and the Grand Coulee." But, the report continued, in addition to "the fish ladders and elevators contemplated, there is a program for artificial propagation set up which may be put into effect if the fish-passing devices fail to meet expectations. No possibilities, either biological or engineering, have been overlooked in devising a means to assure perpetuation of the Columbia River salmon."

In spite of engineering and biological efforts and promises, the salmon runs declined. In the early 1980s, a major effort was undertaken to salvage the fish. The newly formed Northwest Power Planning Council, a regional agency, was supposed to develop a long-range energy plan for the region and to "protect" and "enhance" fish and wildlife affected by hydropower development in the Columbia Basin.

The council's plan to jump-start the salmon runs included constructing bypass facilities at the dams and developing a so-called water budget, a specified amount of water that would be released to help the fish migrate downstream in the spring. In addition, the council declared that some 44,000 miles of rivers and streams should be protected from hydropower development. But despite these endeavors and hundreds of millions of dollars spent to build fish populations during the 1980s, the wild salmon runs remained moribund. The releasing of millions of hatchery-raised fish tended to cause even more problems for the wild fish. It was the failure of these protection efforts that precipitated the deployment of the Endangered Species Act.

One reason for the salmon's continuing decline is

> *The Kwakiutl Indians, who venerated the salmon, addressed their prayers and offerings: "O Supernatural Ones, O Swimmers." Salmon can accelerate faster than a car over short distances, or glide without apparent effort on unseen currents and drifts. Motion is central to their lives. They feed by it, communicate by it, and, one has to suspect, enjoy it too. It is only late in life when their bodies have been deformed by the spawning process that they lose their mastery of water.*
>
> —BRUCE BROWN
> MOUNTAIN IN THE CLOUDS:
> A SEARCH FOR THE WILD SALMON, 1982

that at some dams, completion of bypass facilities has fallen years behind schedule. But a more fundamental problem is the water budget. For one thing, it has been set at a minimum level. In fact, according to the National Marine Fisheries Service, the Planning Council set the Snake River water budget at the Lower Granite Dam at a level below fishery agency and tribal recommendations for *minimum* flows. Furthermore, the water budget for the Snake has been less than half the proposed allocation. The Council doesn't have the power to mandate that its water budget levels be met by the various water-management agencies, among them the U.S. Army Corps of Engineers, the federal Bureau of Reclamation, and the federal Bonneville Power Administration (BPA), which markets about half the power generated in the region.

With the deadline for an NMFS decision on the sockeye petition approaching, there was a regional scramble to head off what was increasingly seen as an inevitable endangered listing. Oregon senator Mark Hatfield helped convene the so-called Salmon Summit, which included representatives of the four northwestern states, the BPA, Native American tribes, environmentalists, fishermen, and utilities.

There is little question that the Salmon Summit was a last-ditch attempt to head off an endangered species listing for the sockeye. Politicians, utilities, and other water users did not want a repeat of the spotted-owl scenario (See Timber chapter). "The only reason the Salmon Summit happened was that the special interests were scared to death," says Andy Kerr, of the Oregon Natural Resources Council. In March 1991, after months of negotiations, summit participants agreed to double the amount of water released from dams on the Snake River in time for that year's spring migration. That amount would still be below the recommended minimum. At any rate, the Army Corps of Engineers, which operates the dams, scotched the measure for 1991, delaying its implementation until 1992. The corps claimed it did not want to make a hasty decision.

In April 1991, the sockeye was proposed for an endangered listing, and the focus on developing a proposal to save the fish shifted much more toward the NMFS. The Salmon Summit, however, continued to meet in hopes of coming up with a plan to influence whatever plan the NMFS came up with.

> *The salmon is the ultimate symbol of the Pacific Northwest.... These stalwarts have fought all the obstacles we've put before them in order to return to the spawning grounds of their birth. We ought to be ashamed of ourselves if we can't save them.*
>
> —GOVERNOR CECIL ANDRUS OF IDAHO
> THE NEW YORK TIMES, APRIL 1, 1991

The Snake River sockeye is only the first of several species of Northwest salmon to be considered for endangered status. Shortly after the Shoshone-Bannock Tribes filed their petition, several conservation organizations and the Idaho and Oregon chapters of the American Fisheries Society petitioned the NMFS to list four other salmon species—the spring, summer, and fall runs of the Snake River chinook, and the coho salmon that spawn in the lower Columbia. The numbers of these fish are also diminishing at a disheartening rate. The run of fall chinooks on the Snake River, for example, has dropped dramatically, from 19,000 in the late 1960s to approximately 600 through most of the 1980s to a mere 315 in 1990, according to the Idaho Department of Fish and Game. In June 1991, the NMFS proposed that the three Snake River Chinook runs be designated as threatened. (It decided against listing the lower Columbia

River coho.) But that may be just the beginning. The Oregon Natural Resources Council, which had already been a petitioner on behalf of the chinook and coho runs, intends to begin filing a number of endangered-species petitions for other West Coast fish in 1991. The council will choose its targets from the roughly 150 species identified as at risk by the American Fisheries Society. "Outside of Alaska, it's hard to find a wild salmon species on the West Coast that isn't having problems," says Kerr.

As for the Snake River sockeye, some observers think that because none were sighted at their spawning grounds in 1990, the species may already be extinct. Only time will tell. Since the fish has a four- to five-year spawning cycle, some could still be out at sea, so whether the species has ceased to exist won't be known conclusively until 1994. We have already lost dozens of fish populations; if the Snake River sockeye and other dwindling species disappear altogether, it will be a sad commentary on the state of the environment in the Northwest. Kerr puts it this way: "Salmon are probably the best indicator of the quality of life in the Pacific Northwest."

WESTERN WASHINGTON WATER WOES

Water problems are not the exclusive domain of salmon, fishermen, and utilities. Western Washington's fast-growing population is beginning to strain the ability of water utilities to deliver enough water to their customers. The Seattle Water Department (SWD), which faces rapidly increasing demands in the coming decades, is singing the water-shortage blues. "Water planners estimate existing supply can keep pace with the demands within the current service area only for the next 5 to 10 years," the department states. Currently, its million-plus customers —the SWD also sells water in the area's fast-growing suburbs—consume an average of 174 million gallons a day, and the department is stretched particularly thin in the summer months, when demand goes up and supplies dwindle. On top of that, the Puget Sound region is expected to grow by 200,000 people over the next 10 years, with daily demand rising to more than 250 million gallons by 2020. So the utility is in the process of putting together a long-range plan to cope with the expected crunch. In the meantime, to encourage conservation, the SWD has added a 50% surcharge on water consumed above a prescribed level during summer and early fall.

The plight of the salmon and the strained resources of the SWD serve as warnings that the era of cheap, apparently unlimited water in the Northwest is drawing to a close. Whether or not that has sunk into the consciousness of the general populace is another question.

WATER QUALITY

The water we use comes from two major sources—surface water and groundwater. Surface water is visible—in lakes, rivers, and streams. Far greater amounts of water are out of sight, in underground reservoirs called aquifers. Aquifers are layers of sand, gravel, or porous rock through which groundwater generally moves extremely slowly. Groundwater may be connected to surface water through underground springs that recharge lakes and streams. Rain and surface water seep down through the earth's crust to replenish aquifers.

Groundwater has become increasingly important as a water source. About half the nation's population

WATER AT HOME

The quality of drinking water is affected by the way it is treated and transported. One source of problems can be chlorine, which is widely employed as a disinfectant in public drinking water systems. Chlorine can combine with organic matter in the water to form a group of suspected cancer-causing chemicals called trihalomethanes (THMs). Following the widespread discovery of THMs in public water systems in the 1970s, the EPA adopted a drinking water standard—or maximum contaminant level—of 100 parts per billion (ppb) for THMs. The standard, however, does not apply to systems serving fewer than 10,000 people. In addition, some experts consider the current 100 ppb level too high to adequately protect public health. In Washington, several systems have either exceeded or come close to the THM standard in the past two or three years. One major system, the Skagit Public Utility District, has consistently recorded THM levels above the standard, including one reading of 347 ppb in May 1990.

Water can also become tainted with toxic substances once it leaves the treatment plant. Lead, a common hazard, is most likely to come from pipes or solder in home plumbing systems. Throughout the early 1900s, lead was commonly used in indoor pipes and in piping between residences and public water mains. Plumbing installed prior to 1930 is most likely to contain lead. After that time, copper pipes largely replaced lead ones, though lead solder continued to be used. However, lead levels in water are likely to be higher in houses that have new copper pipes with lead solder, according to the EPA, since, as plumbing systems age, mineral deposits coat and seal the inside of the pipes. Lead can continue to leach in appreciable quantities if the water is especially acidic.

In 1991, the EPA reduced its drinking water standard of 50 ppb of lead to 15 ppb at the tap. More than 40 million Americans may be using water that contains levels of lead higher than the proposed EPA safety standard. Fetuses, infants, and young children are the most vulnerable to lead's effects, which can include damage to the nervous system, brain, and kidneys. Lead in drinking water may contribute 10% to 20% of the total lead exposure in young children. For infants whose diets are primarily composed of infant formula made with water, lead in tap water may make up 40% to 60% of total exposure to the toxic metal. If you suspect that your water may contain too much lead, get it tested. Tests should range in cost from about $20 to $100. Your local health department can recommend a reputable local laboratory.

In 1986, the federal Safe Drinking Water Act was amended to require the use of "lead-free" pipe, solder, and flux in public water systems or any plumbing or fixture connected to a public water system. "Lead-free," however, means that solders and fluxes may be 0.2% lead, while pipes and pipe fittings may contain up to 8%. Water standing in fixtures made of brass that contains 8% lead can still accumulate "dangerously high levels of lead," Arthur Perler, the former chief of science and technology in the EPA's Office of Drinking Water, maintained in a November 1990 letter to *The New York Times*. "The public must protect itself."

A couple of steps can be taken to reduce lead exposure immediately. The highest lead levels are

> generally found in water that has been in the pipes for more than six hours. So, although it flies in the face of water-conservation practices, the most effective way to safeguard against lead intake is to flush your drinking water taps when you get up in the morning or return home after work, by letting them run until the water becomes as cold as it can get. This can take up to two minutes. (Such flushing may not be effective in high-rise buildings.) Also, use cold water for cooking and drinking; hot water dissolves lead faster than cold and thus is likely to contain higher concentrations of the metal. The best solution is to get the pipes replaced, if possible.

—and more than 90% of rural residents—depends on groundwater for drinking water. In Washington and Oregon, about two-thirds of the population gets its drinking water from underground sources, whereas about 90% of Idaho's drinking water comes from aquifers. More than half the inhabitants of Montana use groundwater for domestic purposes.

Until the 1970s, groundwater wasn't generally thought to be vulnerable to the kind of chemical pollution that affects surface water. The prevailing wisdom was that the layers of soil and rock overlying aquifers provided an adequate defense against chemicals that might leach from above. That perception proved to be sadly incorrect. Not only can chemicals migrate downward in various ways, but once they reach underground water supplies they don't disperse or break down as readily as in surface water. Instead, they usually travel along with the groundwater in slow-moving "plumes" of contamination.

More than 200 chemical substances and other contaminants have been detected in groundwater nationwide, and every state in the country has reported contamination. In the mid-1980s, the Congressional Office of Technology Assessment identified 33 different sources of groundwater contamination. Some are obvious: toxic waste dumps, municipal landfills, industrial sites. But others, not so visible, are far more widespread: underground gasoline and chemical storage tanks, septic tanks, pesticides and fertilizers. Thousands of drinking water wells have been shut down across the country because of contamination from these sources.

Here's a quick glance at some of the problems. According to the Environmental Protection Agency, the nation has some 5 million to 6 million underground storage tanks for gasoline and chemicals. Hundreds of thousands may be leaking their contents into the ground, probably into underground water supplies. About 30,000 hazardous waste sites are considered potential candidates for inclusion on the EPA's Superfund list of toxic waste sites that qualify for federal cleanup dollars. Millions upon millions of tons of pesticides and fertilizers spread across large tracts of land also threaten underlying aquifers. By the late 1980s, about 60 pesticides had been detected in groundwater in 30 states. Even septic systems—and approximately 25% of homes in the United States have them—pose a serious threat to groundwater with the possibility of leaking chemical and biological contaminants.

In some respects, the Northwest is blessed with relatively pristine underground water supplies. The region doesn't have a history of heavy industrialization, as do many other parts of the country, and large areas are not densely populated. Consequently, we have not suffered the problems of, say, New Jersey.

But our aquifers are not immune to the taint of chemicals. Groundwater contamination has been found in 32 of Washington's 39 counties. Because there's no monitoring program that tests all wells in the state for the myriad potential contaminants, the real extent of the problem is unknown, according to the Washington Department of Ecology. Oregon's Department of Environmental Quality says that the number of known groundwater contamination sites has increased over the last few years from a few to hundreds, partly due to stepped-up efforts to identify such contamination.

The relative seriousness of contamination sources depends on an area's prevailing economic activities. Oregon, for instance, cites agriculture as the number one threat to groundwater quality, although industrial activity is blamed for contamination at about 75 sites. The state also reports that groundwater contamination has been found at 23 of 25 monitored solid waste landfills. Montana, by contrast, ranks agriculture low on its list of groundwater threats; it has identified underground storage tanks, injection wells, and septic tanks as its most important problems. As of April 1988, Idaho had logged more than 350 cases of actual or potential groundwater contamination, half of which were the result of leaking underground petroleum storage tanks or petroleum spills. Hazardous material spills accounted for another 102 cases. Idaho also lists feedlots and dairies, landfills, hazardous waste sites, and pesticide use among its top potential sources of groundwater contamination.

Although Washington is the most heavily industrialized and the most densely populated state in the Northwest, the state's Department of Ecology still ranks agricultural activities as the most serious threat

> As of April 1988, Idaho had logged more than 350 cases of actual or potential groundwater contamination.

Water 99

to groundwater. Washington's hazardous waste problem is more extensive than that of the other northwestern states. Of the 741 hazardous waste sites listed by the Department of Ecology, 178 are known to have tainted groundwater, and another 308 are considered potential threats. Nearly half the sites known to be contaminating groundwater are located in densely populated King and Pierce counties. The Evergreen State also has more than 40 federal Superfund sites requiring cleanup of hazardous waste contamination, more than the other three states combined. The most notorious of these is the U.S. Department of Energy's Hanford Nuclear Reservation, in southeastern Washington, where contamination from radioactive and chemical waste sites has spread over 200 square miles and is seeping into the Columbia River.

FROM GROUND TO TAP: YOUR DRINKING WATER

Until the late 1980s, there was no systematic monitoring of drinking water supplies for most chemical contaminants. Under the federal Safe Drinking Water Act (SDWA), public water supplies only required testing for about two dozen substances. In 1986, however, Congress amended the SDWA, mandating that the EPA set health standards for a much greater number of contaminants in drinking water, and calling for states to begin testing public drinking water systems for a wider variety of organic chemicals, including many commonly used in industry or agriculture. In 1988, the Washington Department of Health (DOH), which is responsible for regulating drinking water quality in the state, started monitoring public water systems for 59 organic chemicals.

By early 1991, the DOH had detected chemical contamination in 125 of 2,360 drinking water sources (2,080 of them groundwater). Some 36 different chemicals were found, usually at low levels. According to Jane Ceraso, manager of the DOH's Organics Monitoring Program, volatile organic chemicals (VOCs) have been detected in the greatest number of water supplies in Spokane, Clark, Pierce, and Grant counties.

Oregon's public water supplies are less affected by chemical contamination. By the end of 1990, the state had completed testing on 462 public water systems each serving 200 or more people, and detected VOCs in only 11 systems. According to Kurt Putnam, a field sanitarian with the Oregon Health Division's drinking water section, VOCs have been found in significant concentrations in only three—two systems near Portland, and one serving a development of about a hundred homes called Lakewood Estates, in Marion County.

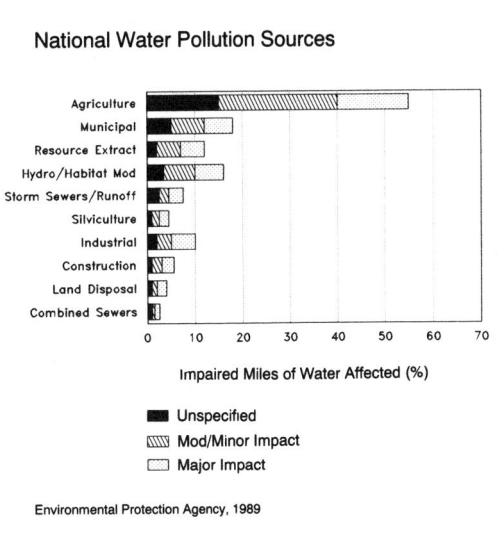

At Lakewood Estates, whose well water was laced with a cancer-causing substance called dichloroethylene, the Health Division informed residents that the levels of that chemical might result in one additional cancer case for every 10,000 persons exposed to it over a lifetime. That's 10 to 100 times higher than the EPA's acceptable risk level for can-

National Water Pollution Sources

Agriculture	
Municipal	
Resource Extract	
Hydro/Habitat Mod	
Storm Sewers/Runoff	
Silviculture	
Industrial	
Construction	
Land Disposal	
Combined Sewers	

Impaired Miles of Water Affected (%)

■ Unspecified
▨ Mod/Minor Impact
□ Major Impact

Environmental Protection Agency, 1989

Water 101

cer-causing chemicals in water. In response, some residents bought bottled water, paid to install filtration systems, or hand-filled containers from a backup well.

The extent of groundwater contamination in the Northwest compares favorably with levels nationwide. According to Ceraso, at least one VOC shows up in about 20% of random samples in nationwide

PERCENTAGE OF STATE AND TERRITORY POPULATIONS SERVED BY GROUNDWATER FOR DOMESTIC SUPPLY

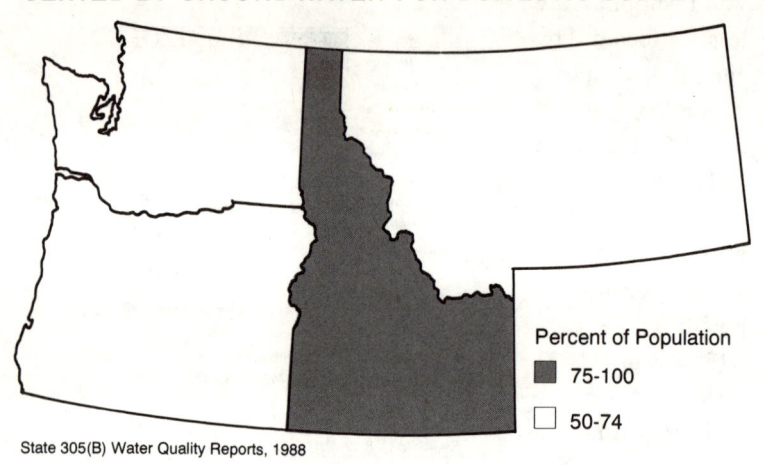

Percent of Population
■ 75-100
☐ 50-74

State 305(B) Water Quality Reports, 1988

surveys of groundwater. Some states have even more severe problems. In Massachusetts, during the first year of VOC monitoring, chemical contamination was detected in nearly half the more than 500 wells tested.

Still, the costs of eliminating chemicals in groundwater can be extremely high, particularly for smaller systems. After the city of Vancouver, Washington, discovered tetrachloroethylene, a suspected carcinogen, in one of its well fields in March 1988, it spent several million dollars to carry clean water to the affected area, to investigate and track the contamination, and to install treatment equipment. The much smaller city of Moses Lake in central Washington spent more than $100,000 just to conduct preliminary investigations into its chemical contamination. That figure doesn't include remediation measures. When it comes to groundwater contamination, "some damage may, in fact, be irreversible," concludes the Washington "Environment 2010" report, a 1989 assessment of the state's environment.

But testing public water supplies is only the first step in assessing the extent of drinking water contamination. In 1993, water suppliers will be required to monitor for the presence of about 60 additional chemicals, many of them pesticides. These requirements extend only to public water systems—those with 15 or more individual hookups or serving more than 25 people year-round. Systems with less than 15 hookups, which provide water to 15% to 20% of the nation's population, are not covered by the federal Safe Drinking Water Act. Contamination in private wells is generally only detected randomly or if the homeowner suspects that a problem exists.

The danger to private wells was illustrated in 1989 when the Washington Department of Ecology announced it had detected pesticide contamination in 23 of 81 wells (67 of them private) in Whatcom, Yakima, and Franklin counties. (The wells tested were selected for possible vulnerability to pesticides. A later resampling confirmed contamination in 20 of them.) Seven wells, all in Whatcom County, contained pesticide levels above those that state or federal health agencies consider safe. One of the wells contained concentrations of potent cancer-causing ethylene dibromide (EDB) 60 times higher than the EPA safety level. (EDB contamination has already led to the shutdown of more than a thousand wells in Florida.) Still, the Ecology Department said the levels should not pose any health concerns. This survey was the first time the state had intensively sampled groundwater for a wide variety of pesticides, even though Washington is a major agricultural state. Denis Erickson, a Department of Ecology hydrogeologist who conducted the study, acknowledges that "we haven't taken a good look in our state" at pesticide contamination of groundwater. Erickson attributes this partly to a lack of funding and partly to the perception, particularly west of the Cascades, that there is plenty of water. But based on the results of the three-county study, he maintains that "we are obligated to look in other areas to see if we have problems."

SURFACE WATER

Maintaining and improving the quality of surface water has been a priority since the early days of the environmental movement. Back in the 1960s, Lake Erie was considered terminally contaminated. Ohio's Cuyahoga River earned a place in environmental infamy by bursting into flames, inspiring the Randy Newman song "Burn On." Closer to home, Lake Washington, east of Seattle, was so choked with foul-smelling algae spawned by the huge volumes of sewage plant effluent pouring into it, that swimming was unsafe.

Serious water pollution helped catalyze the environmental movement of the late 1960s and early 1970s and led to Congress's passage of the Clean Water Act (CWA) in 1972. The CWA had the admirable goal of making the nation's waters "fishable" and "swimmable" by 1983. The EPA and state environmental agencies embarked on a program to control industrial waste and untreated or minimally treated sewage routinely discharged into our rivers, streams, and coastal waters. Under the CWA, industrial facilities and sewage treatment plants that discharge pollutants into surface waters must first obtain permits from the EPA or a state agency under the National Pollutant Discharge Elimination System (NPDES). These permits are subject to renewal every five years. Nearly 50,000 industrial and 15,000 sewage treatment plants have permits.

Between 1972 and 1986, industry and government spent more than $400 billion to control water pollution. Much of the government money went into the construction and upgrading of sewage treatment plants. This was a high priority, not only because of the large volume of domestic sewage, but because municipal wastewater turns out to contain a lot more than what is flushed down the toilet or drains out of the shower: it often gets mixed with toxic substances used in the home. In many municipalities, industrial facilities discharge great quantities of toxic chemicals into the sewer systems.

These expenditures have bought a lot of improvement in surface-water quality. Since 1972, the number of people served by improved "secondary" sewage treatment (which removes higher levels of contaminants) or better has increased by half. The amount of sewage and toxic waste dumped into surface waters has decreased substantially. Fish and vegetation have returned to many bodies of water.

One of the first environmental success stories was the cleaning up of Lake Washington in Seattle. In the late 1950s, 20 million gallons of effluent was pouring into the lake each day. The lake filled with algae, and visibility in what had once been a remarkably clear body of water had fallen to 3 feet by

1963. By 1968, thanks to the efforts of Metro—metropolitan Seattle's sewage authority, created to address water pollution problems—virtually no effluent was emptying into the lake, and its former clarity was returning. By 1977, Lake Washington was clearer than had ever been recorded.

Yet a significant proportion of the nation's surface water remains contaminated. In 1988, 39 states reported that they had enacted fishing restrictions because of high levels of toxic chemicals found in fish. In its latest summary of state data, the EPA reports that 30% of the river and stream miles, 26% of the lake acreage, and 28% of estuarine waters (bodies of water where salt and fresh water mix) assessed for water quality did not fully meet standards for the uses for which they have been designated, such as swimming, fishing, and drinking water supply.

In the Northwest, more than 67% of Washington's nearly 4,900 miles of rivers and 57% of the surface lake area do not meet clean-water goals. About 25% of Montana's river miles and 75% of its lake areas assessed do not fully meet standards. In Oregon, 24% of lake acres and 55% of river miles do not measure up.

Why do surface water quality problems continue? For one thing, EPA and state permits allow industries and municipal treatment plants to discharge certain levels of pollutants. According to figures gathered by the EPA, in 1988 about 360 million pounds of toxic chemicals were discharged nationwide by industrial facilities operating with NPDES permits. Another 570 million pounds of toxic chemicals were sent to public sewage treatment plants. Sometimes the levels allowed by these permits may not be stringent enough to protect water quality, particularly in the case of toxic discharges. In Washington, the state has identified more than a hundred industries as polluters of 45 waterways contaminated with excessive concentrations of toxic chemicals.

One of the major regional dischargers of toxic chemicals is the pulp and paper industry. In September 1990, the EPA warned that 21 pulp and paper mills, including two Weyerhaeuser plants in Washington, might be releasing enough of the highly toxic and suspected cancer-causing chemical dioxin into nearby waters to significantly increase the risk of cancer in people who regularly consume fish caught in those waters. In February 1991, the agency ordered eight pulp mills in Washington, Oregon, and Idaho to cut their discharge of dioxin into the Columbia River Basin to less than 10% of 1988 levels by 1994. (The mills claim they have already cut discharges substantially.) Our neighbors to the north add to the problem: discharge from the Celgar Pulp Company mill in British Columbia is thought to be the source of dioxin and another toxic chemical called furan detected in fish in Lake Roosevelt behind Grand Coulee Dam. The contamination has prompted Washington's Department of Health to warn people that they should limit their consumption of fish caught in the lake.

In the Puget Sound area alone, industries legally disgorge 400 tons of heavy metals and 200 tons of petroleum hydrocarbons annually. The EPA estimates that half of all toxics discharged into the sound are legally released under current permits. Many of these toxic chemicals bind to particles in the water, which then settle in sediment. Concentrations of toxic chemicals in some parts of Puget Sound are a hundred times higher—or more—than in the cleanest rural bays. These high concentrations have been linked to adverse effects on fish, including liver tu-

> In the Puget Sound area alone, industries legally disgorge 400 tons of heavy metals and 200 tons of petroleum hydrocarbons annually.

Water 107

> ### OIL SPILL PREVENTION
>
> In March 1989, the *Exxon Valdez* ran aground in Alaska's Prince William Sound and dumped nearly 11 million gallons of oil into the water. Although it takes a disaster on the scale of this incident to capture worldwide attention, oil spills are unfortunately a relatively routine occurrence. From 1979 to 1990, there were 25 spills of 25,000 gallons or more on the West Coast, according to an October 1990 report by the States/British Columbia Oil Spill Task Force. Half a dozen of those were in or near the waters of Washington and Oregon.
>
> A major fear is that a catastrophic oil spill could hit Puget Sound, through which more ships travel than any other body of water on the West Coast. Nearly one in five of those vessels is an oil tanker.
>
> To try to head off a disastrous spill before it occurs, the Washington legislature passed a measure in 1991 that created an office responsible for preventing oil spills, funded by a tax of 5 cents on each barrel of oil. Among other prerogatives, the safety office has the power to institute a program for vessel inspection and approve plans to control ship traffic. The legislation also requires handlers of oil and hazardous materials to develop spill-prevention plans and mandates that oil shippers carry more insurance.

mors, fin erosion, and reproductive problems.

An added problem is posed by facilities operating with expired permits, permits that, to begin with, were generally less stringent than updated and renewed permits. During the early 1980s, a huge backlog of expired permits bedeviled the understaffed EPA and state regulatory agencies. The backlog had been reduced to about 30% of dischargers by the end of 1988, a large reduction from the early 1980s, but still a significant number.

In addition, many dischargers simply don't meet their permits' requirements. A Washington Public Interest Research Group (WashPIRG) review of regulatory data found that some 33 of the state's 42 major industrial facilities violated their discharge permits during a 30-month period from 1986 through mid-1988. In a rather unflattering assessment of the state's pollution-control efforts, WashPIRG concluded that "compliance with current NPDES permit limits is a bare necessity for even the most minimal attempt to clean up our water."

NONPOINT POLLUTION

Municipal and industrial discharges are known in pollution-regulation jargon as "point sources." That's because the waste comes from one distinct "point," or location. As point sources have been increasingly brought under control, another universe of contamination, known as nonpoint pollution, has come to light. Such contamination is diffuse and difficult to pinpoint and thus to regulate or control. Major categories of nonpoint contamination include agricultural practices—particularly pesticide and fertilizer use, irrigation, grazing, and animal care and feeding—construction, logging, runoff from abandoned and active mines and septic systems, and runoff from urban and suburban areas that does not enter municipal sewer systems. Even air pollution has been shown to settle on surface waters, either through direct fallout or through precipitation.

The major agent behind nonpoint pollution is rain. Rainwater washes sediments and chemicals on the surface of the soil into nearby bodies of water; storm water running over farmlands, for example, can carry pesticide and fertilizer residues with it. Nonpoint pollution also can contain viruses, bacteria, and nutrients, as well as a wide variety of toxic substances such as gasoline, oil, and heavy metals.

Although nonpoint pollution was virtually ignored during the 1970s, it is now recognized as an extremely serious source of surface-water contamination. It is responsible for 65% of the contamination in the nation's "impaired" rivers, 76% in "impaired" lakes, and 45% in environmentally damaged estuaries, according to the EPA. Idaho reported in 1988 that 57% of its surface waters "experience[d] nonpoint source impacts," while only 7% were affected by point sources. Montana maintains that nonpoint sources are to blame in the cases of nearly all state surface waters that "do not fully support their designated uses." In Washington, nonpoint pollution is at least partly responsible for 34 of the 45 waterways' being listed by the state as contaminated with excessive levels of toxic chemicals. Nine commercial shellfish beds in Puget Sound, making up nearly 18% of the sound's commercial shellfish-growing area, have been put on more restricted status since 1986, almost entirely because of nonpoint pollution, according to the Puget Sound Water Quality Authority. The sources of contamination include leaking septic systems, animal waste, and polluted storm water.

Of particular concern in the Northwest is pollution from logging operations. Traditional clear-cut logging techniques lay bare large tracts of land, and because these are often hillsides, heavy rains can wash large amounts of silt into nearby streams and rivers. This not only diminishes water quality in general, but can choke sensitive fish-spawning areas with mud. Many biologists place at least some of the blame for dwindling fish runs in the Northwest on runoff from logging operations. A study by the Idaho Department of Fish and Game, for example, found that populations of westslope cutthroat trout, once present in great numbers throughout the Northern Rockies, now survive only in wilderness and roadless areas. In logged areas, they have all but disappeared. (See Timber chapter.)

Nonpoint pollution is much more difficult and complicated to control than point sources. Thus, most efforts to control nonpoint pollution have focused on voluntary "better management practices" to reduce the amount of contaminants in the runoff, rather than on trying to mandate maximum levels of contamination through a discharge permit system.

CRACKS IN WATER QUALITY ENFORCEMENT

State and federal efforts to control water contamination and provide clean drinking water are still dogged by a lack of resources and personnel. In Washington, for example, an independent consultant found in 1988 that the state's drinking water program faced "a resource crisis, which has greatly reduced the program's effectiveness." The study found that among states of comparable size, Washington had the "highest number of community water systems per full-time employee" and "consistently ranks among the top five states for the number of EPA violations." This is in large part attributable to the great number of water systems, particularly small ones, a number that has doubled in the past decade. Many of the small systems, located in rural and suburban areas, are improperly constructed and haphazardly managed. Many don't regularly comply with testing requirements, either out of ignorance or because of a lack of resources and technical know-how.

But the state and counties have only about a third of the employees needed to monitor all their water systems, according to Bill Liechty, former acting director of the Washington Health Department's

SUPERFUND SITES

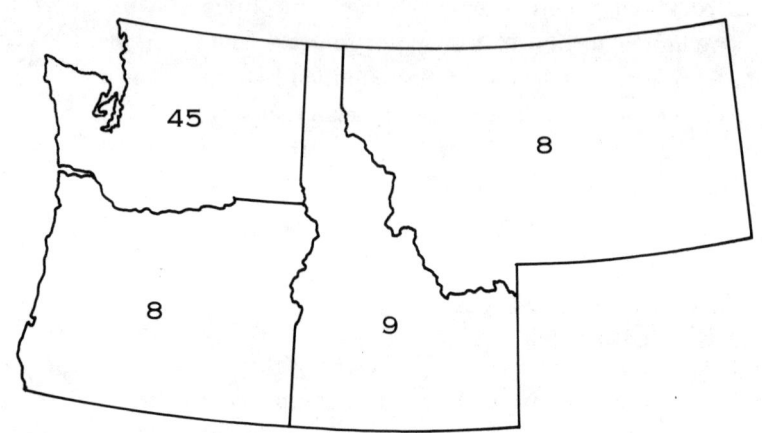

- Washington: 45
- Montana: 8
- Oregon: 8
- Idaho: 9

Environmental Protection Agency, 1991

drinking water division. A 1986 survey indicates that Oregon is relatively better staffed, but resources are stretched even further in Idaho and Montana. Little has changed since the 1988 consultant's report, says Jane Ceraso, of Washington's drinking water program. On top of that, she explains, the work load will only increase in coming years as state regulators are required by amendments to the federal Safe Drinking Water Act to expand their monitoring and enforcement activities.

The Washington Department of Ecology program charged with enforcing compliance among industrial and municipal dischargers also suffers from a lack of resources. A 1990 report by the Washington State Commission for Efficiency and Accountability in Government describes an understaffed, underfunded agency beset by "continuous upset and chaos" and overwhelmed by the magnitude and complexity of the laws it has to implement and the regulations it has to enforce. Among the consequences is a lack of compliance and a rapidly increasing number of expired and unissued water-quality permits, the number of which ballooned from about 350 in 1986 to more than 600 in 1990. In scathing language, the report notes "an erosion of the agency's credibility with permittees and a perception that [the Department of] Ecology does not know what it is doing."

Stating that the "current permit work load cannot be handled by the number of assigned staff," the report estimates that the size of the program and its budget would have to more than double to provide comprehensive regulation of the current 1,050 permitted dischargers. Furthermore, the Department of Ecology has estimated that some 10,000 dischargers should eventually be included in some kind of control program. To accomplish that, the report predicts that the water quality program will need a 500% increase in funding over current levels.

The chances of getting that kind of funding are vanishingly small, especially with many state budgets facing a financial crunch. In 1991, the Washington State Senate, instead of increasing the budget for the water pollution control program, actually cut its funding. "I think everybody's for a clean environment until it comes to paying for it," Rep. Nancy Rust, chairwoman of the House Environmental Affairs Committee, told a reporter.

WHAT YOU CAN DO

Despite the Northwest's reputation as a conservation-minded region, we probably waste water just as much as other Americans. We must develop an awareness that high-quality water is a precious and limited resource; otherwise, voluntary measures to conserve water aren't going to have much effect. There are a number of things individuals can do to save water—and money.

The average resident of a single-family home in the United States uses 77 gallons of water daily. Much of that is wasted. The largest single consumer of water in the home is the toilet, so repairing leaks, avoiding unnecessary flushing, and installing low-flow toilets or devices that reduce the amount of water flushed can save large amounts of water.

Other suggestions for conserving water:

• Take shorter showers or install low-flow shower heads.

• Use washing machines and dishwashers only when you have a full load. According to the Washington Energy Extension Service, clothes washing accounts for 22% of water use in American homes.

• Don't let faucets run unnecessarily when you're brushing teeth, shaving, or washing dishes.

• Avoid using garbage disposals, which can consume more than 11 gallons of water a day.

• Consider reducing the frequency of activities such as washing cars or watering lawns.

Voluntary conservation, however, will only go so far, particularly because household use accounts for a small percentage of overall consumption—11% in Washington in 1985. Farmers and industry can also institute conservation measures. Farmers, for example, can cover or line irrigation ditches or use more efficient sprinkler systems.

Altering the traditional way of pricing water, which in fact encourages consumption, would help spur conservation efforts. According to Washington's "Environment 2010," several municipal utilities and irrigation districts are trying to reform their rate structures.

RESOURCES

Adopt a Beach
PO Box 21486
Seattle, WA 98111-3486
(206) 624-6013

Organizes community stewardship of Washington State waters. Programs include estuary rehabilitation, monitoring of paralytic shellfish poisoning, and beached-bird surveys. Involves schools in the Washington Coastweeks celebration (which includes both educational and restoration activities such as mud-flat safaris and beach cleanups), as well as projects that continue throughout the school year. In 1990, published a comprehensive guide to groups caring for the Puget Sound area and related volunteer opportunities.

Alliance for the Wild Rockies
Box 8731
Missoula, MT 59807
(406) 721-5420

See listing in Growth chapter.

American Fisheries Society
5410 Grosvenor Lane, Suite 110
Bethesda, MD 20814-2199
(301) 897-8616

Association of fisheries personnel, professional fishermen, biologists, state agencies, and others concerned with conservation of fish resources and further research into fish and water quality issues. Comments on National Fisheries and Forest Service regulations and management of resources; local chapters have petitioned to place wild salmon on the endangered species list. Call national office for divisions and state chapters.

The Campaign for Puget Sound
PO Box 2807
Seattle, WA 98111-2807
(206) 382-7007

Created to develop a strategy to safeguard Puget Sound by bringing public pressure to bear on the state of Washington to fund and enforce water quality goals. Intends to serve as a watchdog over government agencies and others who affect the sound's water quality. Startup is scheduled for fall 1991.

Citizen's Clearinghouse for Hazardous Wastes
West Coast Office
PO Box 33124
Riverside, CA 92519
(714) 681-9913

A national organization founded by former Love Canal resident Lois Gibbs, CCHW assists local groups with organizing, media relations, research, and helping citizens deal with experts on an equal footing. In Washington and Oregon, groups are fighting incinerators and working with waste management; contamination episodes are dealt with on an individual basis.

Citizens for Clean Industry
629 Slice Street
Anacortes, WA 98221
(206) 293-6453

Formed to address industrial pollution in Skagit County, this group monitors regulation of the major refineries in the area.

Citizens to Save Puget Sound
PO Box 420
Seahurst, WA 98062
(206) 431-0444

Stopped Metro (the Seattle-area sewage and transit authority) from running a sewage pipeline from Renton to Seahurst Park on the Olympic Peninsula; researched and backed an alternative system that had less environmental impact and was more cost-efficient. Members continue to monitor activities affecting Puget Sound.

Clark Fork Coalition
PO Box 7593
Missoula, MT 59807-7593
(406) 542-0539

A local group in western Montana focusing on issues that concern the quality of the Clark Fork River, including contamination from the nation's largest Superfund site, located at Butte. Has extended its reach into Idaho to address the entire river and other threats to it, including timber cutting, mining operations, and algae growth.

Columbia Basin Fish and Wildlife Authority
Oregon Department of Fish and Wildlife
2501 SW First Avenue, Suite 200
Portland, OR 97201
(503) 326-7031

The only organization of its kind in the United States, this coalition of fish and wildlife agencies (both federal and state, from Washington, Oregon, Idaho, and Montana) and 13 Indian tribes coordinates and implements the fish and wildlife provisions of the Pacific Northwest Power Planning and Conservation Act. Currently working on a proposal to increase the amount of water flowing over the dams in order to double the fish runs. Federal and state funded.

Friends of the Earth
Northwest Office
4512 University Way NE
Seattle, WA 98105
(206) 633-1661

See listing in Growth chapter.

Friends of the San Juans
PO Box 1344
Friday Harbor, WA 98250
(206) 378-2319

See listing in Growth chapter.

Friends of Union Bay
5026 22nd Avenue NE, Suite 2
Seattle, WA 98105
(206) 525-0716

See listing in Growth chapter.

Greenpeace Action
Northwest Office
4649 Sunnyside Avenue N
Seattle, WA 98103
(206) 632-4326

A priority of the regional office is a campaign against pollution from pulp and paper mills, because of the number of mills in the region; the entire organization is involved in a range of environmental issues. Direct action (such as blockading a chlorine shipment in Puget Sound) is done only by staff members; volunteers are welcome for office work.

HANFORD EDUCATION ACTION LEAGUE (HEAL)
1720 N Ash Street
Spokane, WA 99205
(509) 326-3370

A research and education center concerned with the management of radioactive and toxic and chemical waste from the past 45 years of weapons production at the Hanford Nuclear Facility in southeast Washington State. Studies of past radiation releases and their possible health effects are ongoing, as are attempts to block future plutonium production.

IDAHO CITIZEN'S NETWORK
PO Box 1927
Boise, ID 82701
(208) 385-9146

A grassroots membership group organizing around many issues: civil rights, health care for uninsured people, groundwater quality. Each of the five chapters has its own localized issues; the Silver Valley/Kellogg chapter focuses on cleanup at the Bunker Hill Superfund site, contaminated by a hundred years of smelting. One staff member sits on the governor's Ground Water Quality Council, which works on statewide policy. Promotes education and involvement of low- to moderate-income people.

IDAHO CONSERVATION LEAGUE
PO Box 844
Boise, ID 83701
(208) 345-6933

A grassroots group that deals with a constellation of issues, all centered on preserving Idaho's wilderness. Reviews timber sales, mining proposals, and water quality standards, and is working on an antidegradation plan for Idaho's waters. The public lands program, Wild Idaho, concentrates on protection of Idaho's roadless areas; ICL is also working on a federal level to secure more comprehensive wilderness protection for the state. Specific strategies are developed for individual issues, and public education is ongoing. Has met with mixed results, but recent successes include Idaho's 1989 Clean Lakes Act and a state hazardous waste management plan in 1987.

IDAHO RIVERS UNITED
PO Box 633
Boise, ID 83701
(208) 343-7481

Like the Northwest Rivers Council (with which it is associated), Idaho Rivers United works with grassroots groups on rivers issues and does water policy advocacy with state and federal agencies.

MILITARY TOXICS NETWORK
National Toxics Campaign Fund
RR1, Box 2020
Litchfield, ME 04350
(207) 268-4071

This grassroots organization funded by the National Toxics Campaign (based in Boston) promotes "ecojustice" for communities harmed by toxic contamination from federal facilities. The network will send you a Network Starter Kit, a report card for your local military base, and a newsletter. Has the only citizens' toxics lab for providing data other than government information. The Military Toxics Network, headquartered in Maine, attends to problems all over the country.

MONTANA ENVIRONMENTAL INFORMATION CENTER
PO Box 1184
Helena, MT 59624
(406) 443-2520

See listing in Growth chapter.

NATIONAL AUDUBON SOCIETY
Rocky Mountain Regional Office
4150 Darley Avenue, Suite 5
Boulder, CO 80303
(303) 499-0219

See listing in Growth chapter.

NATIONAL WILDLIFE FEDERATION
Northern Rockies Natural Resources Center
240 N Higgins Street
Missoula, MT 59802
(406) 721-6705

See listing in Growth chapter.

Northern Plains Resource Council
419 Stapleton Street
Billings, MT 59101
(406) 248-1154

Founded by ranchers who banded together on national energy policy and coal-mining issues in the early '70s and rising energy prices in the '80s. Still organizes around protecting natural resources from the effects of industry and overuse (hard-rock and coal mining, groundwater protection, sustainable agriculture). Recently, waste issues have gained visibility as other states have sought to establish large landfills in Montana. Offers services to a number of smaller affiliate organizations, giving staff time and organizational materials, and in general working toward the "empowerment of rural people." Recently published *Reclaiming the Wealth: A Citizen's Guide to Hard-Rock Mining in Montana*.

Northwest Coalition for Alternatives to Pesticides
PO Box 1393
Eugene, OR 97440
(503) 344-5044

A policy-influencing organization involving the public through education on the dangers of and alternatives to pesticides. Has been instrumental in designing and lobbying for state groundwater legislation in Oregon, legislation that requires a citizen advisory group to be set up any time nonpoint contamination is found. Also involved in challenging lax pulp mill permits in the Northwest. Publishes the *Journal of Pesticide Reform*.

Northwest Environmental Defense Center
10015 SW Terwilliger Boulevard
Portland, OR 97219
(503) 244-1181 x707

This student-government–funded organization of law students, attorneys, and community members was founded by Lewis and Clark College students 21 years ago. Very involved in fisheries issues, copetitioners for listing several stocks as endangered species. Filed intent to sue Weyerhaeuser over the illegal emissions of a mill in Coos Bay.

Northwest Indian Fisheries Commission
6730 Martin Way E
Olympia, WA 98506
(206) 438-1180

Provides a unified voice for Washington's 20 treaty tribes to express environmental concerns. The tribes, as comanagers of the fishery resources of the state, negotiate with the state government on regulations, data management, and monitoring programs. Developed the Timber, Fish, and Wildlife Agreement, which forms the framework for Washington timber resource planning. Strong proponents of cooperative management—among tribes, environmentalists, and industry—rather than litigation.

NORTHWEST POWER PLANNING COUNCIL
851 SW Sixth Avenue, Suite 1100
Portland, OR 97204
(503) 222-5161, (800) 222-5161 outside Oregon, (800) 452-2324 in Oregon

See listing in Energy chapter.

NORTHWEST RIVERS COUNCIL
4516 University Way NE
Seattle, WA 98105-4511
(206) 547-7886

Helps smaller river-protection groups obtain Wild and Scenic designation—the only preventative measure against hydropower and damming—for free-flowing rivers. Currently actively campaigning for the Wild and Scenic Proposal in Washington, which would get a number of the state's threatened rivers into the program. A similar bill has already passed in Oregon. Public support is mobilized through education and clarification of the designation for landowners. (See also Idaho Rivers United, an affiliated group, and Oregon Rivers Council.)

NORTHWEST STEELHEAD AND SALMON COUNCIL OF TROUT UNLIMITED
PO Box 2137
Olympia, WA 98507
(206) 433-3122

Primarily a conservation organization; main concern is habitat restoration for all cold-water fishes (trout, salmon, steelhead, Dolly Varden, etc.)—in spite of its recreational anglers membership. Offers opportunities for hands-on work on stream and lake rehabilitation, fish raising, and tagging salmon and steelhead. Conducts an extensive education program in area schools.

Oregon Natural Resources Council
522 SW Fifth Avenue, Suite 1050
Portland, OR 97204
(503) 223-9001

Like the Northern Plains Resource Council, this group helps give voice to 50 or so smaller organizations, but strategy runs more to legislation and litigation. Recently particularly aggressive in the area of ancient forest protection, obtaining court injunctions against timber sales and filing endangered species petitions for threatened Northwest salmon runs.

Oregon Rivers Council
PO Box 309
Eugene, OR 97440
(503) 345-0119

Main architects of Oregon's 1988 Wild and Scenic Act; involved in monitoring the now-protected rivers and implementing the act. As with the Northwest Rivers Council, efforts are made to educate landowners on how river conservation can be an economic (as well as natural) boon.

Oregon Trout
PO Box 19540
Portland, OR 97219
(503) 244-2292

Statewide (although work in Columbia River Basin extends into Idaho and Washington) membership organization dedicated to the protection and restoration of native fish and their habitat. Recently were lead petitioners in the drive to have certain wild salmon species put on the endangered species list. Other projects include working with state agencies on policy development and implementation, coordinating tribes around habitat issues, and data collection on native fish populations.

Public Interest Research Group (PIRG)

MontPIRG
360 Corbin Hall
Missoula, MT 59812
(406) 243-2907

OSPIRG
1536 SE 11th Avenue
Portland, OR 97214
(503) 231-4181

WashPIRG
340 15th Avenue E, Suite 350
Seattle, WA 98112
(206) 322-9064

PIRG (Public Interest Research Group) has offices in Montana, Oregon, and Washington, all targeting environmental and consumer issues. Issues emphasized vary by region and change each year depending on urgency. Recent focus has been on pesticide use, pollution prevention, and resource recovery. Does door-to-door canvassing to educate the community and solicit funds. Recent work includes lobbying for a "right to know" ordinance and Superfund cleanup in Montana, and citizen suits against companies that don't comply with EPA standards in Washington. Both OSPIRG and MontPIRG are focusing on recycling issues. OSPIRG recently campaigned for a far-reaching recycling initiative that was defeated by Oregon voters in November 1990. During Montana's recent legislative session, MontPIRG lobbied with other environmental groups for a recycled materials and recycling programs bill; as a result, a task force was set up to determine the role of recycling in Montana's future.

Puget Sound Alliance
10545 41st Place NE
Seattle, WA 98125
(206) 548-9343

Active in the protection of Puget Sound. Activities on water-related issues, watchdog agencies, education programs, and citizen monitoring programs. The alliance's SoundKeeper, a full-time "environmental ombudsman," addresses complaints and problems. Helped lobby to establish the Puget Sound Water Quality Authority. Publishes *Sounder* newsletter.

Selkirk Priest Basin Association
PO Box 181
Coolin, ID 83821-0181
(208) 448-1813

Covers an area from the Canadian border down to the confluence of the Priest and Pend Oreille rivers. Watches water quality and timber issues and works with the Priest Lake ranger district and Idaho Department of Lands. Networking with other groups provides a line into other state issues. Not a lobbying group, but keeps a close eye on relevant legislation.

SIERRA CLUB

Cascade Chapter
1516 Melrose Street
Seattle, WA 98122
(206) 621-1696

Oregon Chapter
c/o John Albrecht
3550 Willamette Street
Eugene, OR 97405
(503) 343-5902

Northern Rockies Chapter
c/o Edwina Allen
1408 Joyce Street
Boise, ID 83706
(208) 344-4565
See listing in Growth chapter.

SIERRA CLUB LEGAL DEFENSE FUND

Northwest Office
216 First Avenue S, Suite 330
Seattle, WA 98104
(206) 343-7340

Rocky Mountain Office
1631 Glenarm Place, Suite 300
Denver, CO 80202
(303) 623-9466
See listing in Growth chapter.

SNAKE RIVER ALLIANCE

PO Box 1731
Boise, ID 83701
(208) 344-9161

Group monitoring the Idaho National Engineering Laboratory seeks responsible solutions for nuclear waste disposal and prevention of further nuclear buildup. Mainly a research and education organization, also works in community organizing and volunteer training (how to lobby, how to testify at a hearing). Successes include the blocking of a proposed "special isotope separator" and limiting the Air Force's plans to expand Idaho's Saylor Creek Bombing Range.

WASHINGTON ENVIRONMENTAL COUNCIL
4516 University Way NE
Seattle, WA 98105
(206) 527-1599

See listing in Growth chapter.

WASHINGTON WILDERNESS COALITION
PO Box 45187
Seattle, WA 98145-0187
(206) 633-1992

See listing in Growth chapter.

WATERWATCH
921 SW Morrison Street, Suite 534
Portland, OR 97205
(503) 295-4039

Main purpose is to promote water policy in Oregon that provides for fish, wildlife, public health, and a sound economy. Is active in litigation, lobbying, and intervention in administrative proceedings. Key issues include stream flow restoration and balancing public and private uses of water resources.

AIR

UNSEEN HAZARDS

THE FORGOTTEN ENVIRONMENTAL ISSUE

THE FALL AND RISE OF AN AIR POLLUTION BILL

A WHOLE NEW SET OF PROBLEMS

THINKING GLOBALLY IN THE NORTHWEST

A CLEAR FUTURE?

AIR

When Congress passed the Clean Air Act in 1970, the country was gasping for breath. Factories freely belched black smoke and smog choked our cities. When the newly created Environmental Protection Agency (EPA) went to work on the problem, it initially focused its efforts on six common pollutants known to be harmful to human health and damaging to crops and vegetation. Those efforts have produced tangible results: industrial plants now release substantially less pollution, new automobiles emit 95% fewer hydrocarbons than did their 1960s counterparts, and emissions of lead, which can cause brain and nerve damage, were cut by 96% from 1970 to 1989, mainly through the reduced use of leaded gasoline. Nationwide, ambient levels of carbon monoxide, airborne particulates, sulfur dioxide, nitrogen oxides, and low-level ozone have also decreased significantly.

There was plenty of concern in the Northwest about air pollution before the Clean Air Act was passed: Oregon established a state air pollution control agency as far back as 1952. And there was good reason to be worried: by the early 1970s, carbon monoxide (CO) readings in downtown Seattle, for example, violated federal standards nearly one out of every three days.

The Northwest has reflected the nation in its success in combating air pollution. Carbon monoxide levels in Washington's atmosphere have decreased by about 30% since 1979. Emissions of sulfur dioxide from pulp and paper mills and fluoride from aluminum smelters have been drastically reduced, and Seattle now exceeds the federal CO standard only one or two days a year.

But after two decades of progress in the clean-air fight, a comprehensive review of the condition of Washington's environment in 1990 called air pollution "the most significant environmental threat" in the state. Other sources of exposure to environmental contaminants "pale by comparison to air pollution," says Kent Swigard of the Puget Sound Air Pollution Control Agency (PSAPCA). On an EPA map of air-quality trends, red and orange splotches indicating areas of unhealthy air are spread out over much of western Washington and Oregon, northern Idaho, and northwestern Montana. The Portland metropolitan area, where about half of Oregon's population resides, appears as out of compliance for CO

AREAS NOT MEETING NATIONAL CARBON MONOXIDE STANDARDS

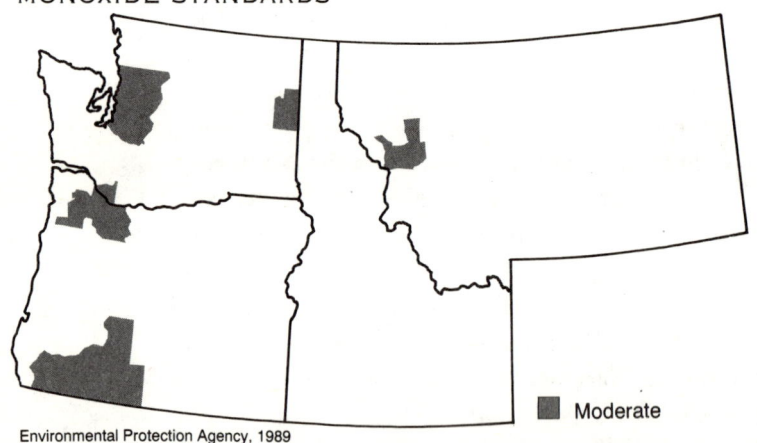

■ Moderate

Environmental Protection Agency, 1989

and ozone. Problems with airborne particulates extend from the Eugene area down through Grants Pass to Medford and over to Klamath Falls.

More than 3 million Washington residents breathe air with potentially unsafe concentrations of pollutants. Greater Spokane is the second-worst area in the nation when it comes to carbon monoxide; only Los

130 Northwest Greenbook

Angeles has higher levels and more days when the federal standard is violated, according to the Washington State Department of Ecology. At least 1.5 million Washington residents inhale unhealthy levels of airborne particulates, which can aggravate respiratory and cardiovascular problems, damage lung tissue, and lead to cancer and early death. The Washington "Environment 2010" assessment says that nearly half a million state residents are "at extreme risk of negative health effects from air pollution." Small particulate pollution alone contributes to the deaths of more than a hundred Washingtonians each year.

What happened? For one thing, despite all the regulatory effort directed at industrial emissions, many facilities continue to release large amounts of pollution. Until recently, Washington was one of only a handful of states that did not require industry to periodically update air pollution control permits. This means that state regulators didn't have an effective way of forcing a firm to upgrade its antipollution technology unless there was a plant modification or expansion. The reason for this regulatory gap, explains Stuart Clark of the state Department of Ecology, was pressure from industry, which "vehemently opposed" such a program. Clark says the result is that many facilities are "operating with [pollution control] technology that is 15 or 20 years old."

But industry is responsible for only 25% of air pollution in Washington, the most heavily industrialized state in the Northwest. (In Oregon that figure drops to 6%.) More culpable, in the aggregate, are the millions of smaller, more diffuse sources—small businesses, gas stations, motor vehicles, wood stoves —that continue to generate a tremendous load of contaminants. Not only are these sources rarely monitored, but their numbers have continued to grow

Air 131

rapidly as population and economic activity have increased. Sheer numbers have overwhelmed what progress has been made, particularly in the case of motor vehicles. Even though a single auto doesn't produce as much pollution as it did in the early 1970s, the use of cars has increased so much over the last couple of decades that it has all but blotted out previous gains. In the Puget Sound area, the number of vehicle miles traveled is increasing four times faster than the population. So-called mobile sources—autos, trucks, motorcycles, etc.—now account for more than half of the air pollution nationwide. In Washington, motor vehicles are now responsible for about 40% of the total air pollution. The Oregon figure is 36%. (Idaho and Montana have not tabulated statewide statistics for air pollution.)

The region's relative abundance of wood is, surprisingly, proving a bane, at least when it comes to air quality. It's ironic that just a few years ago wood was boosted as a safe, renewable fuel (remember those SPLIT WOOD NOT ATOMS bumper stickers?). In fact, burning wood emits hundreds of times the amount of pollution of other forms of heat, namely, natural gas, electricity, and oil. Wood smoke is a major air pollution problem over much of the Northwest. In Oregon, emissions from wood stoves are responsible for nearly 12% of the state's air pollution, and wood smoke is largely responsible for some of the highest airborne particulate levels ever recorded in the country. Washington's million or so wood stoves and other wood-burning devices are responsible for a fifth of the state's air pollution. As much as 85% of the air pollution recorded in some Puget Sound–area neighborhoods during winter months can be traced to wood smoke. The smoke is a particularly serious problem during inversions, when

PSAPCA: Leader of the Pack

The Puget Sound Air Pollution Control Agency (PSAPCA) operates out of nondescript digs in Seattle. But while its acronym and its offices may not be particularly impressive, its record of fighting air pollution is. With a staff of only about 50, PSAPCA polices a four-county area that is home to some 2.7 million people and much of Washington's industry. So far, the agency has compiled an enviable record of protecting public health with far-reaching regulations and innovative programs. "We're willing to be out ahead. We take some heat," says Anita Frankel, PSAPCA's air pollution control officer.

In 1990, the agency came out with regulations to control toxic emissions from industry, in advance of either federal government or state action. It has launched an aggressive public awareness campaign over the dangers of asbestos in home renovations. It also took on the Washington Department of Ecology over what PSAPCA claimed was lax state control of pollution from three pulp plants and an aluminum plant in the Puget Sound area. "Those four plants are operating with technology that's back in the 1970s. It's a bad situation that needs correcting," according to PSAPCA spokesman Kent Swigard.

Under a long-standing agreement, the state is responsible for regulating the politically powerful pulp and aluminum industries. Over the years, however, PSAPCA has grown increasingly frustrated with the state's inability to control those major polluters. Finally, in May 1990, the chief of PSAPCA's compliance division wrote a strongly worded letter telling the ecology department that

Air 133

> its "outdated standards have presented a major obstacle to the protection and enhancement of air quality within PSAPCA's geographic jurisdiction." After some wrangling, Ecology agreed to let PSAPCA conduct joint inspections of the plants.
>
> The agency will have its work cut out for it in the coming years, as it struggles to hold the line against air pollution in the fast-growing Puget Sound area. At least it can take heart in the knowledge that it has a lot of admirers, including some in other regulatory agencies. "They're real cowboys," says one staffer in the EPA's regional office in Seattle. "I love 'em."

the polluted air doesn't blow away. A study by a PSAPCA consultant, using data from Yakima, Washington, found that twice as many people suffering from chronic obstructive pulmonary diseases were admitted to hospitals during a period of stagnant air. "Literally thousands of people are suffering, and nobody offers them any relief," says Donna Larson of Citizens Against Woodstove Fumes. And the problem is not confined to major urban areas. Wood smoke is a major reason that several areas in Montana and Idaho do not meet federal air-quality standards.

Regulators have begun to take on wood smoke pollution. Several years ago, Oregon began a program that limits the amount of airborne particulates that new wood stoves may emit. The EPA then developed a similar nationwide "certified" wood stove program that now bans the sale of new noncatalytic stoves that emit more than 7.5 grams of particulates per hour or catalytic stoves that give off more than 4.1 grams per hour. In Washington, pollution-control agencies can ban the burning of wood and fine violators when air quality becomes unhealthy. (People with no other adequate source of heat are exempt from the bans.) The state has also given PSAPCA the option to ban the use of uncertified stoves—those that do not meet emissions standards—beginning in July 1995, if the agency considers such a measure necessary to protect public health. In the city of Missoula, Montana, and surrounding areas, the installation of new fireplaces is not allowed, and permits are no longer issued to burn wood as the sole source of heat.

Outdoor burning is also an offender, accounting for about 10% of the air pollution in Washington and more than 20% in Oregon. Each year, about 367,000 acres of Oregon's agricultural and forestlands are burned to control infestations of insects and to prepare for replanting and reforestation. Field and forestry burning in Oregon produces about seven times the amount of particulates as are emitted from industrial sources.

Particulate Emissions From Home Heating Devices

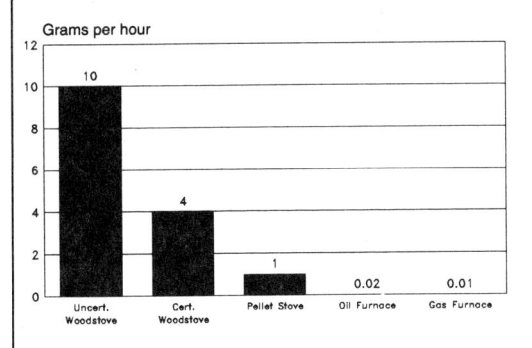

Grams per hour

Source	Grams per hour
Uncert. Woodstove	10
Cert. Woodstove	4
Pellet Stove	1
Oil Furnace	0.02
Gas Furnace	0.01

Puget Sound Air Pollution Control Agency, 1990

UNSEEN HAZARDS

When it comes to air pollution, what you can't see *can* hurt you. Until recently, health officials waging war against ozone and other more visible components of smog did not go after so-called toxic or hazardous air pollutants, many of which are cancer-causing. These pollutants, which industry releases in large amounts, are often invisible but can be deadly. Carcinogenic air pollutants contribute to as many as 1,500 cancer fatalities a year, according to the EPA.

Toxic air pollutants are not a new problem. The 1970 Clean Air Act directed the EPA to regulate more than 300 toxic pollutants. But two decades after the passage of the act, the agency had set standards for a total of only seven. The shortfall, says the American Lung Association, was due to "industry pressure and EPA timidity." A key obstacle to the promulgation of standards was a lack of data, according to Elizabeth Waddell, an air pollution expert with the EPA's regional office in Seattle. Studies on the health effects of relatively small concentrations of air pollutants on large populations are extremely complicated, and the EPA has had to shoulder the burden of proving that pollutants are harmful. "We'd be tied up in court," Waddell explains. "You almost had to show people dying in the streets to a get a change [in the law]."

But the magnitude of the toxics problem hit home when Congress ordered that industry make public the amount of toxic chemicals it releases into the environment. In 1987, the first year for which data are available, industry poured 2.6 billion pounds of toxic chemicals into the nation's air. These emissions were completely legal, and the figures came as a shock to many people, including corporate officials. Some are taking action: the chemical manufacturer Monsanto has pledged to reduce the company's toxic emissions to 10% of their 1987 levels by 1992. American Telephone and Telegraph announced that it will try to eliminate toxic air emissions by the year 2000. Here in the Northwest, the Boeing Company has pledged to cut its hazardous releases in half by 1995.

Boeing's pledge is significant because the aerospace giant, as the region's largest industrial company, accounts for more than 20% of toxic air emissions in Washington State. Though the Northwest is not among the most heavily industrialized regions of the country, industry reported pouring more than 50 million pounds of toxic substances into the air of the four northwestern states in 1988, including suspected cancer-causing trichloroethylene, chloroform, and formaldehyde. Washington and Oregon together account for nearly 90% of the toxic air pollutants released in the region in 1988. The problem is particularly acute in heavily industrialized areas such as Washington's King County, where 7.6 million pounds of toxic air pollutants were released in 1988. Of that total, nearly 1.2 million pounds were potentially cancer-causing.

But once again, large industrial sources are only part of the toxic air-pollutant problem. According to Washington's "Environment 2010," as much as 90% of toxic air contaminants in the state comes from wood stoves, motor vehicles, and small businesses such as gas stations, dry cleaners, and metal-cleaning shops.

THE FORGOTTEN ENVIRONMENTAL ISSUE

The region's air pollution problems have caught many people off guard. "Historically, people have considered this to be the pristine Northwest," observes Stuart Clark of the Washington Department of Ecology. Air pollution is a problem that was supposed to have been solved a long time ago. That attitude translates into a loss of public interest—and

> Industry reported pouring more than 50 million pounds of toxic substances into the air of the four northwestern states in 1988.

Air 137

therefore political clout—in the issue of air quality, once a major environmental concern. Kent Swigard, public information officer for PSAPCA, recalls that in the late 1970s he covered environment as a beat for Spokane's *Spokesman-Review*. When he left the paper, however, no reporter was assigned to the beat. Such neglect contributed to the widespread failure to strengthen air pollution laws, the understaffing of regulatory programs, and the diversion of scarce funds and resources to other concerns, such as cleaning up toxic sites, or water quality programs.

> *It seems to me that Montana is a great splash of grandeur. The scale is huge but not overpowering. The land is rich with grass and color, and the mountains are the kind I would create if mountains were ever put on my agenda."*
>
> —JOHN STEINBECK
> *TRAVELS WITH CHARLEY*, 1962

At the moment, state air pollution control programs are generally hampered by a shortage of staff and funds. The Montana Air Quality Bureau, for example, has about two dozen employees, augmented by a few local health departments. (The bureau expects to add another half-dozen employees starting in July 1991.) The lack of financial resources in Montana can even make relatively simple solutions to serious pollution problems hard to come by. Take the case of Libby, a town of about 3,000 near the Idaho border. Libby suffers from some of the highest levels of airborne particulates recorded in Montana. That's largely because the town is situated in a valley that traps wood smoke and road dust. A vacuum sweeper would probably alleviate the dust problem, but, according to Stan Sternberg of the Montana Department of Health and Environmental Sciences' Air Quality Bureau, both the town and the state are too strapped for funds to buy one.

Resource problems also plague the other states in the Northwest. "All across the board we're understaffed," says Andy Ginsburg of the Oregon Department of Environmental Quality's Air Quality Division. "Oregon has a large backlog in permits that need to be issued, is lagging in collection of air pollution emissions data, and is behind in rule development." The Washington Department of Ecology's air pollution control program is probably that agency's most chronically understaffed and badly funded, receiving barely 6% of the budget. Lack of funds is cited by

AREAS NOT MEETING NATIONAL PARTICULATE MATTER STANDARDS

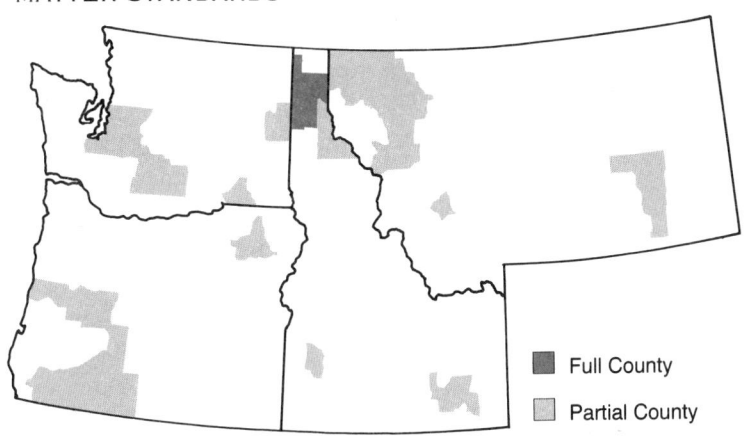

■ Full County
▨ Partial County

Environmental Protection Agency, 1988

the department as the reason for its failure to devise a comprehensive law governing control of toxic air pollutants. The Department of Ecology is supplemented by regional air pollution control agencies such as PSAPCA, which have widespread enforcement responsibilities. But regional agencies are also stretched thin. Currently, PSAPCA is "so understaffed that we are lucky to respond to 10% of the people who call," says Swigard. Currently, the agency has 13 inspectors to check on approximately 3,000 industrial facilities, as well as to try to keep tabs on burn-ban violators.

Regulators' jobs don't promise to get any easier in the future. Sheer growth threatens to overwhelm the admittedly meager defenses against air pollution in some parts of the Northwest. Washington's "Environment 2010" predicts that "population growth will exceed our air pollution controls by 1995. . . . Without new air pollution control strategies, there will be a 30% to 50% increase in carbon monoxide, ozone, and toxic pollutants." In 1991, the Puget Sound area fell out of compliance with federal ozone standards for the first time since 1985. Swigard says, "If we don't do anything more about air pollution, we're going to see our air quality go down the tubes in the next 20 years."

As the 1990s begin, it looks as though air pollution is starting to get the attention it deserves. After years of dickering, Congress finally approved a sweeping revision of the Clean Air Act in late 1990. Among the provisions of the new law:

• By 1996, automobile manufacturers will be required to reduce tailpipe emissions of hydrocarbons, nitrogen oxides, and carbon monoxide.

• The EPA is supposed to set safety levels for emissions of 189 toxic pollutants within 10 years. Facilities in the Northwest that are likely to face controls relatively soon include dry cleaners, metal degreasers, hospital sterilizers, industrial cooling towers, and pulp and paper plants.

• States are required to develop permit programs for industrial polluters that allow the states to recover all direct and indirect program costs from permit fees. This is intended to ensure that states no longer suffer from the financial and resource limitations that have plagued air pollution programs in the past.

The revisions leave many environmental officials in the Northwest feeling that more needs to be done. As Washington ecology department chief Christine Gregoire points out, the Clean Air Act still does not "provide sufficient authority to solve many of our state's specific problems." The new amendments are geared much more toward major industrial areas in the Midwest and the East. Kent Swigard believes that even in the Puget Sound area, the amendments are "not going to have a tremendous impact." They don't address local sources of air pollution such as wood smoke, and, in some areas, they even lag behind local regulations. PSAPCA, for example, already has a rule regulating air toxics.

THE FALL AND RISE OF AN AIR POLLUTION BILL

In an effort to fill in some of these gaps and to improve Washington's air quality, Governor Booth Gardner unveiled for the 1991 state legislative session a plan aimed at attacking the major sources of pollution in the state: motor vehicles, wood stoves, outdoor burning, and industrial emissions. The proposal called for tightening standards for wood stove emissions (although environmental organizations were pushing for even tighter limits), phasing out commercial and residential burning in urban areas by the year 2000, reducing forestry slash burning by 50% over the next decade, and expanding the number of motor vehicles subject to emissions tests. Another provision called for large employers in densely populated counties to develop plans to reduce the number of commuting trips their employees drive alone.

(The trip-reduction provision was later taken out of the clean-air legislation and put into another bill.) Industry would be required to have renewable permits, and would be subject to fees based on emissions volume to help cover the costs of beefing up the regulatory program, as well as increased penalties for violations.

Gardner's plan was an intricately constructed and carefully prepared package designed to distribute costs among all the major sources of air pollution in the state. The idea was that no one group of polluters could claim that it was unfairly singled out, and sabotage the legislation. But the story of the 1991 air-quality proposal shows how even the most carefully crafted bill can almost come unraveled when the legislature gets its hands on it.

The first step for the bill was the House of Representatives, a Democrat-controlled body relatively sympathetic to environmental concerns. Even there, however, various interest groups were able to modify the package. The major changes came in the wood-smoke provisions. Wood stove manufacturers successfully lobbied to ease the emission standards proposed in the original bill. The bill that emerged from the House would reduce the allowable amount of particulates produced by new noncatalytic stoves from the current EPA standard of 7.5 grams per hour to 4.5 grams by 1995, but environmentalists had hoped to slice that level down to 0.5 grams by 1995. Stoves that produce less than 4 grams per hour are already available, points out Linda Tanz, an environmental-health specialist with the American Lung Association of Washington. This new standard doesn't "drive any new technology," she asserts, adding ruefully, "We lost a lot on wood smoke. The wood stove manufacturers did a good job of lobbying."

The other major provisions, however, emerged relatively unscathed. Driven by the hammer of the federal clean-air amendments, industry did not heavily oppose provisions that require five-year renewal and updating of air pollution permits. Outdoor burning provisions were only slightly modified.

But in the Senate, the Environment and Natural Resources Committee, over which conservative rural interests from eastern Washington hold sway, went after the bill with a meat cleaver. The committee voted to reduce the proposed maximum penalties for industry from $10,000 per day for each violation to a flat $10,000, no matter how many violations were incurred. (Federal law calls for $25,000 per day.) It also exempted all certified wood stoves (those producing 7.5 grams or less of particulates per hour) from burn bans and prevented the state from mandating the installation of vapor-recovery devices on gas pumps (scheduled to go into effect in 1992), and on bulk fuel tanks and terminals as well. In a provision that seemed to be aimed directly at King County, which has been considering a phaseout of uncertified wood stoves, the committee appeared to make the Puget Sound area a special target, voting to preempt local government from imposing tighter controls on pollution than state or regional agencies.

The Environment and Natural Resources Committee's actions provoked an immediate and widespread outcry. The press was extremely critical of the committee's gutting of the legislation. Local, state, and federal regulators condemned the revised bill. "They've gone way over the line," PSAPCA's air pollution control officer Anita Frankel told the *Seattle Post-Intelligencer*. EPA officials warned that many of the changes would probably violate the new federal Clean Air Act, and that the EPA would probably have to take over the program.

In the end, the committee's frontal assault on the legislation backfired. "Some of those amendments [from the committee] were so inept, they probably didn't understand the ramifications," says Stuart Clark of the Department of Ecology. The full Senate ultimately undid practically all the weakening amendments the committee had slipped in. Still, the tortuous route the legislation took shows that environment legislation can't be taken for granted, no matter how popular pro-environment rhetoric is.

The Oregon legislature is considering a somewhat different approach to controlling air pollution. One bill introduced in the 1991 session would impose fees on a range of pollution sources, from wood and field burning to automobiles. The fees would then be used to develop less-polluting alternatives. For example, revenues from a fee on a cord of firewood would go to help low-income households switch from wood heat, or at least install cleaner-burning, certified wood stoves. Fees imposed on outdoor burning would promote alternatives such as composting. Automobile fees would be used to promote alternative fuels and traffic-reduction measures.

The Oregon approach illustrates the effort to control air pollution by emphasizing "market-based" approaches rather than relying strictly on further regulation. Economic incentives are receiving more attention among environmentalists and government officials because of the limitations of the traditional method of controlling air pollution through regulations and fines. The latter approach requires more resources and staffing, and quickly runs into the law of diminishing returns. It's hard to modify people's life-styles with regulations. Nor is it possible for air pollution–control agencies, with their limited staffs, to ensure that hundreds of thousands of wood stove owners are complying with burn bans.

But if a combination of financial incentives and voluntary compliance isn't successful in noticeably improving air quality, government agencies may be forced to proceed further down the road of regulation. That's already happening in California, where terrible air pollution problems have necessitated extreme regulatory measures. California's Air Resources Board in late 1990 mandated a reduction in air emissions from power tools such as lawn mowers, leaf blowers, chain saws, and hedge trimmers. These "utility" machines, which produce far more pollution per horsepower than motor vehicles, contribute an estimated 5% of the hydrocarbons and 4% of the carbon monoxide emissions in the state.

While the Northwest has more leeway than southern California, there is little question that more regulation is on its way. The alternative—doing nothing—would be worse. "Do people want to face what's going on in southern California?" asks King County Executive Tim Hill. "We don't want to get to that point." However, regulations must be combined with innovative programs that get to the source of air pollution problems. For government, that means supporting such measures as the development of good, workable public transportation systems and land-use policies that discourage sprawl. For industry, it means trying to reduce emissions and coming up with alternatives to the use of toxic chemicals. None of this will be cheap. The cost of the equipment required by cars to meet more stringent pollution requirements could add $500 to the price of a new automobile. Annual fees collected in Washington under the state air-quality bill could amount to

The Health Effects of Common Air Pollutants

Carbon Monoxide (CO): Inhaled, carbon monoxide enters the bloodstream and interferes with the delivery of oxygen to the body's organs and tissues. Exposure to unhealthy levels of carbon monoxide is linked to impairment of visual perception, learning ability, manual dexterity, and performance of complex tasks. The danger is increased for people who suffer from cardiovascular disease. Transportation sources are responsible for about two-thirds of the nation's CO emissions.

Airborne Particulates: Airborne particulates are thought to affect breathing, aggravate respiratory and cardiovascular problems, and alter the body's defenses against cancer. Those likely to be most sensitive to particulates include people with chronic obstructive pulmonary or cardiovascular disease, the elderly, and children. Particulates come from a variety of sources: burning of wood and other materials, industrial emissions, motor vehicles, and windblown dust.

Ozone (O_3): Ozone is a key component of what is commonly referred to as smog. It is formed when nitrogen oxides react with volatile organic compounds in the presence of sunlight and heat. A distinction must be made between low-altitude ozone, which is an air pollutant, and ozone in the upper atmosphere, which actually provides protection from the sun's harmful ultraviolet radiation. Low-altitude ozone affects not only people with respiratory problems, but healthy individuals as well. Exposure to relatively low levels of ozone for only a few hours has been shown to reduce lung function in healthy persons during exercise;

chronic exposure may cause more long-lasting lung damage. Ozone also adversely affects plants and is blamed for several billion dollars' worth of crop loss each year in the United States. In the Northwest, windblown ozone is a threat to foliage in our national parks and national forests. The rain-drenched climate of the western part of the region makes the ozone problem less acute than it might otherwise be. Still, in 1991, the Puget Sound area fell out of compliance with the federal ozone standard for the first time in six years.

Nitrogen Dioxide (NO_2): Nitrogen oxides can irritate the lungs and reduce resistance to respiratory infections. Los Angeles is the only urban area in the nation that has recorded violations of the annual NO_2 standard in the past decade.

Sulfur dioxide (SO_2): Exposure to high levels of sulfur dioxide can alter the lung's defenses and aggravate respiratory and cardiovascular problems. People most sensitive to SO_2 include those with chronic lung or cardiovascular disease. Both NO_2 and SO_2 contribute to acid rain.

Lead: Lead can damage the kidneys, nervous system, and blood-forming organs. Heavy exposure can cause seizures, mental retardation, and behavioral and learning disorders. Infants and young children are particularly susceptible to the effects of lead exposure. Levels of lead in the nation's air have dropped dramatically since the 1970s, primarily because of a reduction in the use of leaded gasoline.

$20 million. Still, it seems worth the price if we want to maintain our air quality in the face of increased growth.

A WHOLE NEW SET OF PROBLEMS

In the early 1970s, when the battle against air pollution was in its infancy, there was scant recognition that man-made emissions could cause long-term alterations of the atmosphere. Now, global warming, depletion of the ozone layer, and acid rain are considered the most serious—and most intractable—environmental problems, not only in this country but across much of the earth.

One of the first warnings of atmospheric damage came from two scientists, F. Sherwood Rowland and Mario Molina, who postulated in 1974 that a group of chemicals called chlorofluorocarbons (CFCs) were destroying the stratospheric ozone level. Ozone molecules, comprising three oxygen atoms, have radically different effects on humans and other life depending on where they are in the atmosphere. Low-altitude ozone is the harmful pollutant that hangs over our cities, but ozone in the upper atmosphere is extremely beneficial because it helps screen out potentially damaging ultraviolet radiation from the sun. Exposure to ultraviolet radiation can cause skin cancer, cataracts, and other health problems.

Rowland and Molina's theory proved, unfortunately, to be right on the mark. In the mid-1980s, scientists found an ozone "hole" over the Antarctic. More bad news followed: it turned out that the ozone layer over the Northern Hemisphere was also being depleted. The worst was yet to come, however. In April 1991, the EPA, after analyzing data from satellite measurements, announced that the ozone layer over the United States had shrunk by 4% to 5%—which would mean that it was disappearing at about twice the rate previously thought. "It is stunning information," EPA administrator William Reilly told *The New York Times*. "It is unexpected, it is disturbing, and it possesses implications we have not yet had time to fully explore."

CFCs, which are commonly used in air conditioners and refrigerators, react in the atmosphere to become highly efficient destroyers of ozone. After they are released, CFCs eventually migrate to the stratosphere, where they interact with ozone molecules. One chlorine atom from a CFC molecule can destroy up to 100,000 molecules of ozone. A number of other chemicals, including some solvents used in industrial processes, also deplete the ozone layer.

More trouble is occurring at lower altitudes, in what's called the troposphere (the layer that extends up to 10 miles above the earth's surface). In that layer, other chemicals, including carbon dioxide, methane, and CFCs, threaten to warm up the global climate. This they would do by trapping heat that would normally escape. It's currently estimated that the earth's temperature could rise between 2 and 5 degrees centigrade by the middle of the next century. That's enough to profoundly change regional climates and ecosystems. Although no one can accurately catalog the effects of global warming, one likely consequence would be a partial melting of polar ice caps and a subsequent rise in ocean levels. Prime agricultural areas might shift northward, and there are predictions that the midwestern part of this country would become more arid.

The major culprit in global warming (or the greenhouse effect, as it is popularly called) is carbon dioxide (CO_2). Atmospheric concentrations of CO_2 have increased about 25% since the mid-1800s and about 10% since 1958. Carbon dioxide is mainly produced by burning fuels such as coal, wood, and oil. Another contributor to global warming is deforestation, which reduces the amount of vegetation that absorbs CO_2.

Global warming is almost certain to have drastic effects on the Northwest's sensitive ecological systems. The region's shorelines and wetlands could suffer severe damage from rising sea levels, and our snowpack might shrink. Melting snowpack could adversely affect the fish population; Duane Neitzel, a researcher at Battelle Pacific Northwest Laboratories, points out that changes in river flows may have a huge impact on the life cycles of salmon and other fish.

THINKING GLOBALLY IN THE NORTHWEST

Because climate changes are global in scale and often relatively slow, it's much harder to detect any direct cause and effect in terms of local or regional measures. Also, the Northwest's overall contribution to these changes is relatively small in global terms. North America is responsible for about 25% of the world's production of carbon dioxide. Washington, the most populous northwestern state, produces approximately 1% of that portion. That doesn't mean we can ignore the problem, though. Ultimately, everyone contributes to global air pollution. Any time we drive a car, heat our homes with oil or natural gas, or fire up the wood stove we produce carbon dioxide. Reducing CO_2 emissions is a bit like recycling: one person's actions, by themselves, aren't going to have any noticeable effect, but collectively they can make a difference. Because the production of carbon dioxide results primarily from burning fossil fuels, the best way to cut down is to conserve energy whenever possible. To an extent, we've done that since the energy crisis of the 1970s. According to Richard Watson, head of the Washington State Energy Office, due largely to improvements in energy efficiency implemented since 1973, Washington is producing about 30 million fewer tons of CO_2 annually.

Transportation accounts for about half of Washington's CO_2 production, compared with less than a third nationwide. The average fuel efficiency for cars in Washington State is 20 miles per gallon. Replacing current automobiles with ones that get 30 MPG would cut automobile CO_2 emissions by 30%. Carpooling and taking public transportation would obviously reduce that amount even more. Keeping engines tuned and tires properly inflated helps by improving fuel efficiency.

Any measure that reduces the use of electricity or fossil fuel helps. That includes weatherizing homes, buying more efficient appliances, and replacing incandescent light bulbs with more efficient, longer-lasting fluorescent bulbs. The good news here is that not only are most of these measures relatively painless, but they almost always end up saving money over the long term.

But the public can only do so much—we must also have the support of industry and government. If vehicle manufacturers would increase auto fuel efficiency standards gradually to 45 MPG by the mid-1990s (which is technically possible), it would reduce gas consumption by 24%—and thus reduce carbon dioxide production. But they've resisted doing so. During the 1970s and early 1980s, Congress pushed automakers to manufacture more fuel-efficient models, but by the start of the 1990s, the average fuel efficiency of automobiles was actually declining for the first time since the mid-1970s.

Air 151

We also need better public transportation systems and incentives to carpool, if people are going to be expected to relinquish their cars to any significant degree. Unrestricted growth and the proliferation of roads will continue to lead to suburban and exurban sprawl, making a mockery of attempts to encourage people to use public transportation or drive less.

A CLEAR FUTURE?

After years of neglect, the Northwest is beginning to pay more attention to air-quality problems. But continued progress in combating air pollution is not a sure thing. Despite widespread publicity over deteriorating air quality, mobilizing public concern about the problem has been difficult. Perhaps that's because people rarely feel the effects of air pollution immediately, and even visible smog is not as dramatic as, say, a clear-cut. As the attempts to weaken the clean-air bill in the Washington legislature show, there is no consensus on how to strengthen our air pollution laws—or on who should foot the bill. But health officials are sending out the message that even if air pollution regulations are greatly strengthened right now, air quality in some parts of the Northwest will still deteriorate further before it gets better. Delaying action on air pollution only seems to ensure that future costs will increase.

RESOURCES

AMERICAN LUNG ASSOCIATION

1111 S Orchard Street, Suite 245
Boise, ID 83705
(208) 344-6567

825 Helena Avenue
Helena, MT 59601
(406) 442-6556

1776 SW Madison Street, Suite 200
Portland, OR 97205-1798
(503) 224-5145

2625 Third Avenue
Seattle, WA 98121
(206) 441-5100

Working with a coalition of local environmental groups to strengthen air pollution regulations. Follows the air pollution issue closely, monitors government activities, serves on advisory committees, and lobbies.

CITIZENS AGAINST WOODSTOVE FUMES (CAWF)
PO Box 1442
Bellevue, WA 98009
(206) 546-8711

Educates the public about adverse health effects of wood-burning stoves and works with a coalition of environmentalists on pollution issues. Testifies at hearings and is involved in drafting legislation and regulations.

PUGET SOUND AIR POLLUTION CONTROL AUTHORITY
200 W Mercer Street, Room 205
Seattle, WA 98119-3958
(206) 296-7330

See sidebar in this chapter.

Sierra Club

Cascade Chapter
1516 Melrose Street
Seattle, WA 98122
(206) 621-1696

Montana Chapter
c/o James Conner
78 Konley Drive
Kalispell, MT 59901
(406) 752-8925 or contact the field office in Sheridan, WY, at (307)672-0425.

Northern Rockies Chapter
c/o Edwina Allen
1408 Joyce Street
Boise, ID 83706
(208) 344-4565

Oregon Chapter
c/o John Albrecht
3550 Willamette Street
Eugene, OR 97405
(503) 343-5902

See listing in Growth chapter.

ENERGY

CONSERVATION

POLITICS AND REGULATION

ENERGY

The mothballed nuclear power plant near Satsop, west of Olympia, Washington, stands as a kind of monument to a bygone era in the Northwest. It is one of five Washington Public Power Supply System (WPPSS) nuclear power plants begun in the 1970s to bridge the projected gap between the Northwest's electrical supply and its needs. Billions upon billions of dollars after the inception of the scheme, only one of the plants feeds the region's power grid. Two others (including the one near Satsop) are only partly completed, their future in limbo. The last two were abandoned in the early stages of construction. The WPPSS experience turned out to be a boondoggle that resulted in this country's largest public bond default.

Regional energy planners wanted to build the nuclear plants because they were convinced the demand for electricity would continue to grow by leaps and bounds and the Northwest would be desperately short of power by the 1980s. It didn't turn out that way. The recession of the early 1980s actually cut power demand. In the middle of the decade, the Northwest generated more power than it used, and sold much of the excess to California.

In the latter part of the 1980s, however, the trend reversed again. The surplus of a few years before pulled a quick disappearing act, eaten up by burgeoning demand for electricity to feed the region's economic boom and expanding population. As the 1990s begin, the Northwest finds itself in balance: demand about equals supply. The region now uses an average of about 18,000 megawatts of electricity. (A megawatt is the equivalent of 1,000 kilowatts.)

But this happy situation won't continue much

Energy 157

longer. Even with a modest rate of growth, the Northwest is going to need to acquire more electrical power in the next few years or employ what it has more efficiently, making some hard choices about which uses will get priority. As the WPPSS debacle showed, however, predicting future electricity needs is a risky business; how much power the region will need depends on how fast the region's population increases and how fast the economy grows. To deal with these variables, the Northwest Power Planning Council (NWPPC), which is responsible for developing a plan to meet the future energy needs of Washington, Oregon, Idaho, and western Montana, has come up with projections of future energy needs for several different growth scenarios.

If the economy grows quickly, we will need to add as much as 13,000 megawatts to our production capacity over the next two decades—that's about 20 times the amount of electricity the city of Portland currently needs. If, on the other hand, the region goes into a recession, the need for additional power in the future will be cut drastically. So where is needed power going to come from? Developing new sources of electricity is not an environmentally benign activity. Nuclear power not only creates environmental problems, particularly in terms of the disposal of radioactive waste, but carries a heavy stigma stemming from the WPPSS scandal. Even though the two partly constructed nuclear plants could still be completed, as a solution to the region's future energy needs, nuclear power, with its high price tag and unresolved safety questions, is sure to generate massive public opposition.

Another possible source is coal-fired power plants. But these also cause big pollution problems, especially through carbon dioxide emissions, which contribute to global warming. Developing more hydroelectricity means building more dams, and the Northwest is pretty much dammed up already. There are dozens of proposals to build small dams, many of them on rivers and streams in western Washington. But even if these small projects overcome environmental concerns and pass regulatory muster, they will not make much of a dent in the region's overall power appetite. And in fact, the region is facing pressure to reduce its reliance on hydroelectricity in order to ensure the recovery of endangered salmon runs on the Columbia and Snake rivers. (See Water chapter.)

CONSERVATION

What other energy options are there? Well there's cogeneration, a process in which industrially produced heat is used to generate electricity. Although cogeneration sounds great, it is largely dependent on the expansion or construction of industrial facilities by private industry. The NWPPC wants to acquire 650 megawatts of cogeneration power by the end of the century.

Renewable energy sources, such as wind and geothermal energy, are also under consideration. K. C. Golden, executive director of the Northwest Conservation Action Coalition (NCAC), a consortium of citizen groups, environmental organizations, and electric utilities, thinks that the region could eventually pull 8,000 megawatts out of the renewable-energy hat. The NWPPC agrees that there are promising renewable sources but is concerned over uncertain costs and availability.

That leaves conservation. Remember conservation—bundling up in sweaters and turning down the

Energy 159

POLITICS AND REGULATION

State and federal environmental regulations are presumably formulated after consideration of scientific evidence by appropriate officials. These officials also weigh economic considerations, however, which means that industry is not likely to be required to remove every last bit of contamination, since the costs would be prohibitively high.

In the real world, political considerations and influence also enter the regulation equation. Industry and environmental groups lobby in Congress and state legislatures to modify environmental legislation. They also try to exert influence over the regulators themselves. Much of this activity is hidden from public view. Sometimes it breaks out into the open.

That happened here in the Northwest when the Environmental Protection Agency's inspector general reported in January 1990 that Robie Russell, the former EPA regional administrator for Washington, Idaho, Oregon, and Alaska, "took extraordinary steps to prevent formal enforcement actions from being initiated against the owners of the Bunker Hill Superfund site." That site consists of 21 square miles of heavily contaminated soil surrounding a smelter complex in northern Idaho. "Anecdotally, we were told that in some areas the lead level in the soil is so high that it could be mined," the EPA report stated. As a result of Russell's actions, "the smelter complex was allowed to deteriorate to the point that it was declared a public health hazard."

Several months later, the inspector general again hit the Northwest's regional EPA management for "questionable" actions in 10 of 11 cases

> investigated. In one instance, Russell approved an incomplete list of heavily polluted rivers in Idaho, overriding his staff's recommendation that it not be approved. This delayed the implementation of efforts to control water pollution, the report stated. Russell also intervened to reduce penalties against violators of asbestos demolition and renovation standards.
>
> The inspector general's office found that "an atmosphere of distrust and divisiveness existed in the Region." EPA staff also told investigators that they stopped expressing their views on controversial environmental issues, and that "staff recommendations were discounted when outside pressure was applied to environmental issues." Russell resigned in early 1990. He was replaced by Dana Rasmussen, an attorney from Oregon.

thermostat? The Planning Council and conservation advocates hurry to point out that conservation doesn't meaning shivering in the dark. "We're not talking about more sweaters and less light," says Golden. Conservation really means using the electricity we generate more efficiently. Reducing power loss in the electrical transmission and distribution system, weatherizing, building new homes with higher energy-conservation standards, installing more efficient appliances, and using light bulbs that require less electricity are some of the measures that would result in energy savings. Aggressively developing conservation could mean the difference between not having to develop and pay for other, more expensive and environmentally damaging energy sources, such as coal or nuclear.

The council has already decided that conservation is the cheapest, fastest, and most flexible way to meet the Northwest's energy needs, at least over the next decade. In spring 1991, it approved a plan calling for the region to take conservation measures to save about 1,500 megawatts annually. Even conservation isn't going to be cheap, though. The council estimates that the cost of achieving the conservation goals in its plan will be about $7 billion by the year 2000, and, in the words of Ed Sheets, the NWPPC's executive director, these efforts will include "homes, businesses, factories, and farms across the Northwest." That's undoubtedly a bargain, however, when measured against the cost of building nuclear and coal-fired power plants, particularly considering environmental costs.

The NWPPC says the region could obtain 4,600 megawatts of conservation over the next 20 years. The Northwest Conservation Act Coalition, while generally supportive of the plan, maintains that conservation can ultimately provide more than 6,000 megawatts.

But conservation savings aren't going to be realized overnight. For one thing, private utilities may not be gung ho on promoting conservation, since their revenues—and profits—depend on selling, not saving, electricity. "The more kilowatts they sell, the more money they make," K. C. Golden says. But that system of establishing electrical rates may be changing. In April 1991, the Washington Utilities and Transportation Commission approved a plan to allow Puget Sound Power & Light, a fast-growing utility that serves large areas of western Washington, to base its income largely on the number of customers it serves, rather than only on the amount of electricity it sells.

Conservation alone may not be enough to feed the region's energy appetite if growth is too rapid.

There is a wild card in the region's energy mix. This is the effect of recovery plans for endangered Columbia and Snake river salmon. Utilities worry that they are going to lose thousands of precious megawatts of hydropower because of the endangered salmon. But Jim Lazar, a Washington economist, thinks there may be ways of protecting fish while avoiding massive brownouts. In a January 1991 report to the Pacific States Marine Fisheries Commission, Lazar advocated switching about 1,500 megawatts of residential electric space and water heating to natural gas, negotiating seasonal exchanges of power with California and the Southwest, and giving the federal Bonneville Power Administration, which provides about half the region's power, the right to cut off electricity to so-called direct service industries (DSIs)—industrial users, like aluminum smelters, that get large amounts of power at discount rates.

The power exchange makes sense because the Southwest needs more power during the summer months to run its air conditioners, while the Northwest's peak electrical demand—fueled by the need to heat homes and offices—occurs during the winter months. Lazar's proposal to allow the interruption of power to the DSIs is sure to generate heat. Those industries use about 3,500 megawatts (nearly 20% of the region's total). About 90% of that share goes to the energy-intensive aluminum industry. Under long-term contracts, negotiated when there was a power surplus, the DSIs pay extremely low rates for power. Lazar advocates modifying this preferential treatment to the point that direct service industries' power can be completely curtailed if the region needs it for other uses.

The DSIs, not surprisingly, are upset at the prospect of having their supply of cheap power tinkered with. "The [NWPPC] should plan on meeting the industries' current load, plus reasonable growth," John Carr, executive director of the Portland-based Direct Service Industries Inc., said in a 1990 interview. If that power is not available, the not-so-veiled threat is that those companies would simply shut down their operations. That would mean the loss of up to 10,000 jobs.

For the moment, conservation and energy exchanges will help buy time to make the hard decisions that are likely to become necessary as the region continues to grow. Still, it is clear that the era when northwesterners didn't have to think about where their next kilowatt was coming from are over. "Suddenly, the stakes are high and rising," the NCAC noted in its 1990 "Model Action Plan" for the Pacific Northwest. "The question is whether we can rise to the challenge before a crisis pushes us to the brink of costly overreaction." All around us, the indications are that the time for making hard decisions is coming sooner rather than later.

RESOURCES

ALTERNATIVE ENERGY RESEARCH ORGANIZATION (AERO)
44 N Last Chance Gulch, Suite 9
Helena, MT 59601
(406) 443-7272

Action committees explore renewable energy sources as well as sustainable agriculture and recycling. Members receive the quarterly *Sun Times* and can participate in hands-on projects.

AMERICAN COUNCIL FOR AN ENERGY-EFFICIENT ECONOMY
1001 Connecticut Avenue NW
Washington, DC 20036
(202) 429-8873

Publishes brochures "The Most Energy Efficient Appliances" and "Saving Energy and Money with Home Appliances."

FRIENDS OF THE EARTH
Northwest Office
4512 University Way NE
Seattle, WA 98105
(206) 633-1661

See listing in Growth chapter.

NORTHERN PLAINS RESOURCE COUNCIL
419 Stapleton Street
Billings, MT 59101
(406) 248-1154

See listing in Water chapter.

NORTHWEST CONSERVATION ACT COALITION
6532 Phinney Avenue N, Suite 15
Seattle, WA 98103
(206) 784-4585

A coalition of 53 organizations from Washington, Oregon, Idaho, Montana, and British Columbia, including consumer groups, low-income-advocacy groups, environmental groups, and progressive public utilities companies.

Formed in 1981, NCAC is an advocacy group for conservation and clean, affordable energy. Its studies and energy plans are presented to local governments and power companies. Most active in energy policy planning for the whole region. Demonstrates energy-efficient projects and alternatives.

NORTHWEST POWER PLANNING COUNCIL
851 SW Sixth Avenue, Suite 1100
Portland, OR 97204
(503) 222-5161, (800) 222-3355 outside of Oregon, (800) 452-2324 in Oregon.

Created by an act of Congress, the council is mandated to develop a regional energy plan for Oregon, Washington, Idaho, and Montana that provides a "reliable" energy supply at the least cost. It is also required to come up with a program to protect and enhance the fisheries in the Columbia River Basin.

SIERRA CLUB
Montana Chapter
c/o James Conner
78 Konley Drive
Kalispell, MT 59901
(406) 752-8925 or contact the field office in Sheridan, WY, at (307) 672-0425.

See listing in Growth chapter.

RECYCLING

BEYOND CURBSIDE

WHAT YOU CAN DO

RECYCLING

It's a chilly, windy Saturday in December. Clouds scud across the sky over Lake Washington as lanky Robert Blumenthal bounds around the lot at Kirkland Recycling, greeting an intermittent stream of customers, weighing aluminum cans, heaving newspapers onto an ever-growing stack, sweeping up broken glass near the bins. His energy never wanes, even as the afternoon lengthens toward dusk.

Blumenthal is known in the recycling business as a "buybacker." He purchases newspapers, aluminum cans, and other recyclable materials mainly from local residents, and sells them to brokers. It takes someone as indefatigable as Blumenthal to stay in this line of work; his margins are at best razor-thin, even though he has cut his operation to the bone. Recycling has always been a tough business, but never more so than now. Ironically, after two decades of serving as the public's main link to recycling, buybackers like Blumenthal, who gave up a career with the financial firm of Merrill Lynch, are being pushed out of the way by the large waste companies' control of successful municipal curbside collection programs, such as those in Seattle, Bellevue, and dozens of other municipalities in the Northwest.

In 1990, about 100,000 tons of tin cans, plastic bottles, glass, aluminum cans, cardboard, yard waste, newspapers, and other types of paper were collected in Seattle. Programs like this helped Washington State divert 29% of its waste from the landfill or the incinerator that year. Welcome to the new world of recycling, which had, until the last few years, largely been viewed as the realm of the Robert Blumenthals, people who got involved in recycling primarily out of environmental concern.

Recycling 169

Manufacturing products out of recycled rather than virgin materials generally requires less energy. Recycling conserves natural resources such as trees and produces less air and water pollution, in some cases to a dramatic extent. Replacing virgin paper fibers in the paper manufacturing process with recycled or "secondary" fibers reduces energy use by 23% to 74%, air pollution by 74%, water pollution by 35%, and water use by 58%. Studies have also shown that recycling creates more jobs than incinerating waste or placing it in landfills.

Despite all the lip service paid to recycling, the actual proportion of our waste that was recycled had barely climbed above 10% by the mid-1980s. Convenience won out, and mainstream America continued to take its trash to the landfill—around 150 million tons a year. In only a few years, however, that situation has change radically. The U.S. Environmental Protection Agency (EPA) now wants 25% of the nation's waste recycled by 1992, and collection programs for recyclable material have sprung up across the country, with many states announcing even more ambitious goals than the EPA's. One in six Americans now has access to curbside collection programs, and politicians, solid-waste officials, corporate officers, and waste haulers are jumping on the

The Truth About Trash

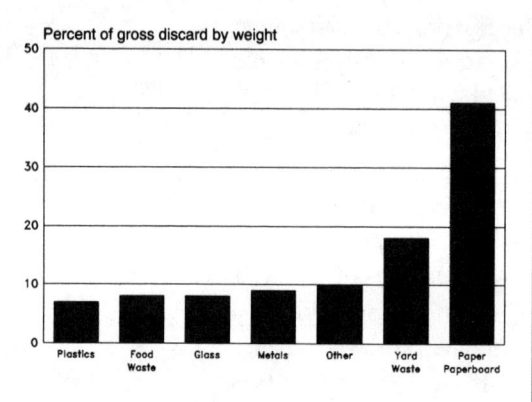

Percent of gross discard by weight

Plastics	Food Waste	Glass	Metals	Other	Yard Waste	Paper Paperboard
~7	~8	~8	~9	~10	~18	~40

Office of Solid Waste, USEPA, 1986

recycling bandwagon so fast that it's in danger of breaking an axle.

What happened? The Garbage Crisis. Our Throwaway Society has reached the limit of its ability to dump all its trash, at least in heavily populated areas. The country is producing so much garbage that we are running out of places to put it all. The cost of waste disposal has suddenly skyrocketed, while landfills, many of them full and contaminating groundwater, are closing down. By 1988, two-thirds of the 20,000 landfills in the nation that had been operating a decade earlier were closed. Another 2,000 are expected to shut down by 1993. Americans have become familiar with images of a garbage barge from Long Island wandering the seas in search of a place to dump its rotting cargo.

But this didn't happen overnight. The amount of garbage Americans—and northwesterners—produce had grown by leaps and bounds in recent decades, but few public officials seemed to take notice. In 1960, Americans produced 88 million tons of waste. By 1988, that figure was nearly 180 million tons—about 4 pounds per person per day, according to the Environmental Protection Agency. By the turn of the century, the EPA expects the nation to produce 216 million tons—a further 20% increase. Amounts vary from state to state: Idahoans, for example, each toss away nearly 6 pounds of trash daily—about 50% higher than the national average.

Washington and Oregon have a head start on most states when it comes to recycling. Washington was already recycling 29% of its waste in 1989, the highest rate in the nation. (Oregon, which has had a major recycling law on the books since 1983, was second, recycling 22%.) Certain specific locales do even better. Seattle has achieved a 36% recycling

Recycling 171

rate, and that's only the start: the city wants to recycle 60% by 1998.

But such success isn't uniform. Montana recycles only 5% of its waste. In Idaho the figure is around 1%. With small populations and large land areas, these two states have not felt the effects of the garbage crisis the way their more populous western neighbors have. Additionally, sparsely populated rural areas can't be expected to generate the amounts of recyclable materials necessary to offset the relatively large collection and transportation costs that recycling programs often entail.

As impressively as Washington and Oregon have performed, they can hardly afford to sit on their laurels. In the coming years, even with recycling, the amount of garbage the Northwest generates will probably increase. In Washington, state residents will still need to toss out more stuff in 2010 than they do now, even if the state achieves its ambitious 50% recycling goal. Much of that increase will simply be the result of population growth. But additionally, the amount of waste generated per person has actually been growing. The solid-waste stream in the Portland area, for example, grew more than 50% from 1984 to 1988, offsetting the city's 26% recycling rate.

Increasing the rate of recycling significantly will require two major steps. first, the scope of collection programs will have to be expanded. Even Seattle's curbside program, frequently lauded as the best in the nation, doesn't include people living in apartment buildings of more than four units. They total 90,000 units, not an insignificant slice of a city whose population numbers about half a million. The city is in the process of extending the service to include apartment building dwellers, beginning late 1991.

But the problem isn't in gathering the stuff. In ▷

WHAT YOU CAN DO

Individual consumers can do a lot to cut down on the amount of waste they generate, even in the absence of wider regulatory measures. The following suggestions and information are a starting point for reducing waste. The organizations listed at the end of this section can provide more detailed suggestions and information.

REDUCTION AND REUSE

The most effective path to waste reduction—as well as to conserving resources and curbing pollution—is simply to reduce the amount of materials used. Avoid excessive packaging—it accounts for nearly a third of what we throw out. When you go shopping, bring a canvas bag. Buy items in bulk, if possible; it's usually cheaper and saves packaging. Avoid disposable items.

Your choices may seem insignificant, but in the aggregate they add up. According to *The Recycler's Handbook* (Earth Works Press, 1990), if every American household avoided using new paper bags for just one trip to the grocery store, we would save as many as 60,000 trees.

Americans receive a staggering 4 million tons of junk mail a year. Cut down on your junk mail (and aggravation) by writing the Direct Marketing Association, Attn: Mail Preference Service, 6 East 43 Street, New York, NY 10017, to request that your name not be given out to any additional solicitors.

Reuse items whenever you can. Aluminum foil can often be used more than once. Plastic tubs are good for storage. Brown paper bags can be used to wrap packages, or reused for shopping.

RECYCLING

An estimated 80% of the waste we produce is potentially recyclable. Realistically, we won't achieve that figure anytime soon, mainly due to a lack of markets and means. But more of our waste can be recycled than currently is.

Paper: According to the EPA, about 40% of our waste stream is composed of paper. Currently 25% to 30% of paper is recycled. Not all paper is created equal, however. Newspapers, which alone make up 8% of waste, are relatively easy to recycle. The problem comes with so-called mixed paper, a sort of catchall category that includes books, junk mail, notebook paper, and scrap paper. Mixed paper tends to be of low quality and have limited markets. Some recycling programs, such as Seattle's, do accept mixed paper and magazines, but this material brings in little money, and much of it currently goes overseas.

Aluminum: Aluminum is a big recycling success story; in 1989, Americans recycled about 60% of the 80 billion aluminum cans they used. The primary reason is that recycled aluminum has a high value. Virtually all recycling centers and collection programs accept aluminum cans. Many also accept scrap aluminum and aluminum foil, but often request that it be separated from cans.

Glass: Glass makes up 7% (by weight) of a typical municipal waste stream. Clear, brown, and green glass beverage containers are all recyclable. Glass recycling has proven so popular, in fact, that in parts of the Northwest there is currently a glut of green glass. But a large portion of the glass in our waste load is "other" glass—e.g., light bulbs, windows, car windshields, beverage glasses. Glass in these forms is extremely difficult or impossible to recycle, often because it is mixed with other materials.

Metals: Metal accounts for 8.5% of waste. Most municipal curbside collection programs pick up "tin" cans—actually 99 percent steel, with a tin coating to prevent rust. Recycling centers sometimes accept metal cans, but generally don't pay for them. The problem is that recycled steel doesn't have a high resale value. Scrap dealers will often take larger amounts of scrap metals such as steel, brass, or copper.

Organic Waste: Approximately 1 out of 6 pounds of what we throw out each year is yard waste. Food waste makes up another 7% of the waste stream. Much of this can be composted. If you don't have any place to put a compost pile, see if your town or city has a municipal compost program.

Plastics: In 1988, Americans discarded nearly 15 million tons of plastics—an increase of almost 500% since 1970. Plastics now account for 8% of the weight and 20% of the volume of our waste stream. Unfortunately, recycling plastic has proven extremely difficult to do; only about 1% was recycled in 1988. The exception seems to be polyethylene terephthalate (PET), which is used to make soft-drink bottles. PET bottles—with the help of state bottle bills—now account for most of the plastic recycled in this country. Some curbside programs, such as Seattle's, accept PET. There is also a small market for high-density polyethylene, or HDPE, which is used to make many types of plastic containers, everything from milk jugs and butter tubs to shampoo and detergent bottles. A few locales, including Seattle, have drop-off sites for HDPE containers. (There is no residential curbside pickup for HDPE in Seattle.)

> Many attempts to recycle plastics have run into problems. In Oregon, for example, plastic recyclers have recently cut down on the number of drop-off boxes. And Seattle found out how problematic the market for recycled plastics could be when Thailand, a buyer of plastics, imposed an import duty making it too costly to ship plastics there.
>
> *Tires:* Every year, Americans throw out 250 million tires, about 80% of which end up in landfills or in tire piles. These piles sometimes catch fire, releasing huge amounts of toxic chemicals into the environment. Currently, only a fraction of the tires in the country are recycled, usually as retreads. At the present time, consumers can't do much to change the situation, except to buy longer-lasting tires, maintain them properly, and turn them in to dealers who recycle.

▷ the 1980s, communities learned how to do that. The second step in the process is figuring out what to do with it—creating new markets to absorb the material collected. The 1990s has to be the decade of creating demand for recyclables, "or we won't be recycling," says David Dougherty, assistant director of the Washington Department of Trade and Economic Development.

The importance of developing demand for products made from recycled, or "postconsumer," material was underscored by a November 1990 Washington State report on creating recycling markets, which projected that the amount of recyclable material recovered statewide will jump to 3.5 million tons in 1995 from 1.5 million tons in 1988, and then to 4.6 million tons in 2010, if state recycling goals are met.

"This material must be marketed as recyclable or it will revert to the disposal stream," the report states flatly.

Already, the amount of material collected is far outstripping the ability or willingness of manufacturers to use it. Prices for recycled newspaper have gone through the floor. Mixed paper brings in almost no money at all. A lot of it ends up in Asia because there is no domestic market for it.

At the Owens-Brockway plant in Portland, a mountain of green glass, estimated to contain more than 30 million bottles, continues to grow because of a lack of buyers. Most of the green glass in this country comes from imported beer and wine bottles, but American bottlers don't use much green glass, and glass colors can't be mixed in the remanufacturing process. Hence, what amounts to an above-ground landfill rises at Owens-Brockway.

Already, the glut of recovered material has cooled some of the waste collection companies' enthusiasm for recycling. Rabanco, a waste collector in Seattle, has complained that the curbside programs are money losers because the glut has driven prices down. The lesson of all this is that without the creation of demand for recyclable products, it doesn't matter how much is collected. The president of the Fort Howard Corporation, a major wastepaper recycler, summed up the situation before a gathering of recyclers in August 1990: "Cities may have enjoyed successful separation and collection programs, but things fell apart when they discovered that no markets exist for their collected [recyclables]. The ultimate indignity is sending those collected materials to the landfill. And believe me, that has happened over and over again."

The demand for some recyclables should increase

in the coming years. In Washington State, 22 new projects that can consume a total of a million tons of recycled materials are planned or under way. The glut of old newspapers, for example, is largely the result of a lack of de-inking capacity, a necessary step in recycling newsprint. So several de-inking projects are scheduled for start-up in the Northwest soon, including a $300 million facility at Weyerhaeuser's Longview, Washington, paper plant. That plant will have the capacity to handle more newspapers than are recycled in all of Washington State.

Paper companies have been reluctant until recently to develop the de-inking capacity necessary for recycling, both because it was expensive and because they were worried about not having a steady supply of material. Also, many paper manufacturers traditionally had a bias against using recycled newsprint because they owned timber that served as the source of virgin material. Now, however, several local and state governments have laws requiring that newspaper publishers use recycled newsprint or face fines and taxes. California, for example, passed legislation mandating that by the year 2000, half the newsprint used must contain 40% recycled fibers. The California newsprint content requirement is, in turn, a major factor be-

Seattle Curbside Recycling Monthly Tonnage Collected

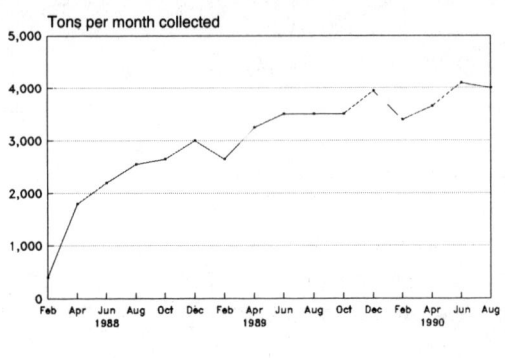

Seattle Solid Waste Utility, 1980

hind the construction of de-inking facilities in Washington.

BEYOND CURBSIDE

Unfortunately, even successful curbside collection programs are not going to solve our waste problems. A more far-reaching solution is to reduce the amount of garbage that we produce in the first place. (See Green Products chapter.)

Waste reduction and reuse are generally more efficient than recycling. Waste reduction, which is the most effective way to conserve, means using less material to begin with. Reuse, the idea behind returnable bottles, consumes less energy than collection and remanufacture. Reuse is not a new idea; it would be accurate to call it a recycled concept. In 1947, for example, nearly 100% of beverage bottles were returnable. But the development of modern manufacturing techniques and materials assured the ascendancy of no-deposit, no-return.

Since the early 1970s, nine states, including Oregon, have passed some form of bottle bill to control litter and to encourage recycling and reuse of bottles and cans. Generally, these programs employ a deposit as an incentive to return containers. They have proven extremely effective: Oregon's redemption rate is more than 90%. A January 1991 report by the congressional General Accounting Office found that the nine states with deposit laws, constituting 18% of the population, were responsible for nearly two-thirds of all the glass and 98% of the polyethylene terephthalate (PET) plastic bottles recycled nationwide. The most stringent law is in Maine, where residents are required to pay deposits on almost all nondairy beverage containers of a gallon capacity

> Waste reduction, which is the most effective way to conserve, means using less material to begin with.

or less.

But do bottle bills undermine curbside collection programs? Not according to the General Accounting Office, which found that curbside recycling and deposit legislation "complement each other and should be seen as compatible tools for managing solid waste." One obvious area in which bottle bills could bolster residential curbside programs is beverages consumed outside the home. These amount to 25% of soft drinks sold, according to the National Soft Drink Association. "The bottle bill states that have 25% recycling goals should realize from one-sixth to a quarter of their overall recycling goal through implementation of the bottle bill," writes Patricia Franklin, director of the National Container Recycling Coalition.

Supporters of beverage container deposits are attempting to get laws passed in 1991 in more than a dozen states and in Congress as well. But those bills face extremely tough sledding. In fact, no state has succeeded in passing a container deposit law since the early 1980s. Opposition can be traced largely to heavy, well-financed lobbying by the packaging and beverage industries. Industry opposition to bottle bills represents a fundamental roadblock to reuse and reduction. Packagers and bottlers want to maintain the throwaway status quo because it enables them to pass the costs of packaging and waste disposal on to consumers. Nearly 10% of the money we pay for food and beverages goes to packaging—$28 billion in 1986. That packing then has to be disposed of. In 1988, containers and packaging accounted for more than 30%—nearly 57 million tons—of municipal waste.

Packagers don't pay the cost of getting rid of that mammoth amount of garbage; they pass it on to businesses, consumers, and taxpayers. Curbside collection programs aren't any threat to packagers because the residents end up paying the costs of collection. Such programs may actually help deflect attention and energy away from alternatives such as bottle bills or restrictions on packaging. In fact, one newsletter reported in January 1991 that the National Soft Drink Association had donated more than $365,000 to 16 state soft-drink organizations to "promote curbside recycling collection and statewide recycling legislation, and to fight container deposit legislation."

Oregon is the only northwestern state with a beverage container deposit law. Efforts to pass a bottle bill in Washington have gone down to repeated defeat. In 1982, a ballot initiative collapsed in the face of opposition from supermarkets, packagers, bottlers, and breweries. A bottle bill has been introduced several times in the Washington State legislature—so far without success. Jules James, who has almost single-handedly fought for a state bottle bill over the past few years, vows to try again in the 1992 session.

Beyond deposit legislation, manufacturers and packagers have also opposed efforts to force them to cut down on wasteful packaging. In 1990, they joined forces to bury a recycling initiative on the Oregon ballot under a blizzard of cash. Among the biggest contributors against the initiative were the American Paper Institute, Procter & Gamble, and a variety of plastics and chemical manufacturers. The far-reaching initiative would have forced packaging in Oregon to meet at least one of the following criteria: be reusable at least five times; be made of at least 50% recycled material; or be manufactured with materials recycled at a rate of at least 15% in the state (a figure to increase to 60% by 2002). Supporters claimed the initiative would have reduced the state's waste stream by up to 800,000 tons a year—about 40% of the garbage produced by Oregonians.

The measure was clearly aimed largely at plastics, which have become an increasingly large portion of the waste stream. Plastics are particularly problematic because they are generally difficult to recycle or reuse. Proponents of the initiative hoped to spur creation of a market for recycled plastics, as well as to reduce excessive plastic packaging. And the battle is not over: supporters carried a streamlined version of the ballot measure into the state legislature in 1991.

In Washington, the issue of controlling packaging has not yet become as publicly contentious as in Oregon. But a state Packaging Task Force, primarily comprising industry members, has put together a plan that is supposed to reduce the amount of packaging 20% by 1998. The task force's recommendations include a requirement that postconsumer material be used in packaging, that plastics contain a code identifying the type of plastic resin used, and that the use in packaging of most toxic heavy metals, such as lead, be banned. Critics—mainly from environmental organizations and local government—say the 20% recycling goal is too low. They are particularly opposed to the task force's recommendation to prohibit local bans on certain types of packaging. Local governments in Washington are already prevented from placing bans on packaging until July 1, 1993, to give industry time to reduce the volume, weight, and toxicity of its packaging. (The city of Port Townsend on the Olympic Peninsula has already prohibited plastic foam containers.)

But industry lobbyists have already killed the recommendation that postconsumer material be used in packaging. The only significant reforms that made it through the Washington legislature in 1991 were provisions mandating coding of plastics and banning the addition of heavy metals to packaging. The legislature's failure to take stronger steps to reduce packaging waste "will likely result in local banning or taxing ordinances that promise to be sweeping," predicts Kitty Gillespie of the state Department of Ecology's Office of Waste Reduction.

State and local governments are going to be major arenas for initiating programs and legislation that promote recycling and waste reduction. At least half a dozen such bills were introduced in the Oregon legislature alone in 1991. How effective those reforms are will be crucial. The 1990s could well decide whether recycling and reuse become viable ways of handling our waste problems.

While each of us can do a lot to make our lifestyles less wasteful, there is a limit to how effective individual actions can be. When so much packaging is made out of nonrecyclable material and without markets for recyclables, it doesn't matter how much material is collected. The creation of such markets and cutting down on wasteful packaging are already becoming major issues, and state and local governments have major roles to play by promoting the use of recycled and recyclable material. In some cases it is going to be a battle, as the Oregon recycling initiative shows. The public can play a major role by showing that it will buy recycled or recyclable goods, and by getting involved politically to change the ways in which our current economic system encourages wasteful production techniques.

RESOURCES

FRIENDS OF RECYCLING
Seattle Solid Waste Utility
505 Dexter Horton Building
710 Second Avenue
Seattle, WA 98104
(206) 684-7666

Sets up volunteer neighborhood contacts to facilitate recycling. Commitment averages about two hours per person per month; responsibilities include relaying feedback on the project from the community to the program manager.

IDAHO WASTE REDUCTION ASSISTANCE PROGRAM
Department of Health and Welfare
1410 N Hilton Street
Boise, ID 83706
(208) 525-WRAP or (208) 334-6664

IWRAP is a waste-reduction and recycling program sponsored by the state. It provides technical assistance to businesses and individuals to reduce waste production and encourage alternative waste management.

KING COUNTY SOLID WASTE DIVISION
450 King County Administration Building
500 Fourth Avenue
Seattle, WA 98104
(206) 296-4468

Publishes the *King County Home Waste Guide*, which offers recycling tips and solutions to current household disposal problems, as well as listing recycling resources in King County.

NATIONAL CONTAINER RECYCLING COALITION
712 G Street, Suite 1
Washington, DC 20003
(202) 543-9449

A coalition of organizations that researches and promotes the benefits of beverage container deposits. The group is active in lobbying for national container deposit legislation.

Northern Plains Resource Council
419 Stapleton Street
Billings, MT 59101
(406) 248-1154

See listing in Water chapter.

Public Interest Research Group (PIRG)

MontPIRG
360 Corbin Hall
Missoula, MT 59812
(406) 243-2907

OSPIRG
1536 SE 11th Avenue
Portland, OR 97214
(503) 231-4181

See listing in Water chapter.

Sierra Club
Northern Rockies Chapter
c/o Edwina Allen
1408 Joyce Street
Boise, ID 83706
(208) 344-4565

See listing in Growth chapter.

Washington Citizens for Recycling
216 First Avenue S, Suite 360
Seattle, WA 98104
(206) 343-5171

This 11-year-old grassroots group tries to promote markets for recycled materials, as well as waste reduction, and has been active in making recycling an issue with local government as well as the public. Works through education (poster distribution in public schools, for example) and outreach activities with industry, government, and the public. Currently working on a five-county program to recycle motor oil.

WASHINGTON STATE RECYCLING ASSOCIATION
203 E Fourth Avenue
Olympia, WA 98501
(206)352-8737

Serves as the trade association for recyclers in Washington State. Promotes recycling, provides education and technical assistance to both members and nonmembers, fosters communication among recyclers and with the public. Membership includes buybackers, brokers, and makers of products that contain recycled material.

WASTE INFORMATION NETWORK
322 W Ewing Street
Seattle, WA 98119
(206)684-2330

Network of small businesses, environmental groups, and state agencies that meets monthly to discuss better waste management for businesses and to outline options for them. Open to the public.

WASTE REDUCTION, RECYCLING, AND LITTER CONTROL PROGRAM
Washington Department of Ecology
4407 Woodview Drive
Olympia, WA 98504-8711
(206)438-7541

GREEN PRODUCTS

**THE SEARCH
FOR STANDARDS**

GREEN PRODUCTS

One day in 1990, David Dougherty, of the Washington Department of Trade and Economic Development, went into a Seattle dry cleaner to pick up some laundry. When he received his clothes, he noticed that the clear plastic wrapping covering them bore a label proclaiming, WE LOVE THE EARTH. THIS PLASTIC IS RECYCLABLE. Dougherty didn't have any way of recycling the plastic, so he asked if the dry cleaner did. He got back a blank stare and the following explanation: "Our buyer saw it [the wrapping] and thought it was pretty." Dougherty took the wrapping home and threw it away. "In fact it isn't recyclable, because there isn't any way to recycle it," he explains.

Dougherty's tale is illustrative of a problem that increasingly confronts consumers: how to sort through a blizzard of environmental marketing claims. What products are *really* the least damaging to the environment? Is a product really biodegradable? Does the label list all the potentially harmful ingredients? It's a problem that threatens to become increasingly complex, as the so-called green marketing craze catches on in business circles and advertising campaigns.

The opportunity to turn green into gold seems irresistible. According to the Marketing Intelligence Service, a research firm based in Naples, New York, 751 products were introduced in 1990 boasting of environmentally beneficial attributes, nearly triple the number in the previous year. A September 1989 Gallup Poll showed that more than 87% of respondents said they would pay more for "environmentally safe" products or packaging. All of a sudden, supermarket shelves are crowded with goods that are "degradable," "ozone friendly," or "recyclable."

But determining the veracity of these claims can be a dizzying prospect. "I know that many consumers who want to do the right thing are completely confused," says Philip Dickey of the Washington Toxics Coalition, a Seattle-based organization trying to reduce the use of hazardous chemicals.

Take Dougherty's plastic wrapper. Perhaps it is technically possible to recycle it. But because neither he nor the dry cleaner has any practical way of recycling it, the claim becomes more than meaningless—it borders on misleading. The problem of deceitful green advertising has become so pervasive that it has caught the attention of state regulators across the country. "Unfortunately, attempts to take advantage of consumer interest in the environment have led to a growing number of environmental claims that are trivial, confusing, or even misleading," asserts "The Green Report," a 1990 study of environmental advertising claims prepared by 10 state attorneys general (including Washington's Ken Eikenberry).

In fact, several manufacturers have had to backpedal after drawing fire for environmental claims. In June 1990, seven states, including Washington, filed suit against the Mobil Chemical Company, a division of the Mobil Oil Corporation, alleging that the firm falsely claimed its "Hefty" trash bags would degrade when exposed to sunlight and therefore would break down when disposed of in landfills. Critics of Mobil's claims pointed out that these bags are generally not exposed to sunlight when tossed into landfills and therefore wouldn't degrade for years. By mid-1991, Mobil had reached out-of-court settlements with the plaintiffs.

In October 1990, American Enviro Products, under pressure from several state attorneys general, agreed not to claim that its "Bunnies" disposable diapers were "degradable" or "biodegradable." And Willamette Industries of Portland, Oregon, a major paper-bag maker, stopped using the words "biodegradable" and "recyclable" on bags used in supermarkets in 17 western states after it was pointed out that most consumers would not recycle the bags and that it would generally be years before the bags broke down in landfills.

Researchers at Cornell University blew a big hole in biodegradable claims when they found that about a dozen plastics used in consumer products carrying a "biodegradable" label did not decompose. "People have jumped to the conclusion that biodegradable plastics are good, but the truth is, they don't biodegrade," one of the researchers told *The New York Times*. Nancy Wolf of the Environmental Action Coalition added, "It's been a marketing ploy whose time has run out."

Some environmental claims, while technically not incorrect, may simply be downright silly. "The Green Report" cites as an example the claim that a polystyrene foam cup helps "preserve our trees and forests." While it is true that polystyrene cups aren't made of paper, the petroleum products they are made of are extremely difficult to get rid of in any way other than tossing them into a landfill.

The Washington Toxics Coalition's Dickey has identified several techniques used to make products appear more environmentally palatable. One approach involves listing with great specificity what the product does *not* contain, but not what it does. For example, some aerosol spray products made without the ozone-depleting chlorofluorocarbons are advertised as "ozone friendly"; but they may contain other ingredients, such as 1,1,1 trichloroethane, that harm the ozone layer. Another technique, which Dickey calls "fantasy ingredients," consists in giving misleading descriptions or definitions of ingredients to make them seem more benign than they really are. An example of this would be describing hydrogen peroxide as a blend of "natural and abundant hydrogen and oxygen." A third method involves combining a less hazardous ingredient with a much more toxic one—for example, saying that a product contains lavender oil when it also contains paradichlorobenzene, a toxic, volatile organic chemical. "Consumers are susceptible to this kind of advertising," Dickey says, "which exploits their . . . lack of technical knowledge."

A consumer backlash may result if people start to feel that companies are simply trying to exploit environmental concerns. There is some evidence that this is happening already. A Gallup survey released early in 1991 found that only 8% of those polled were "very confident" about the accuracy of environmental claims, while nearly half said they were "not confident at all." As a result, "The Green Report" warns, consumers may "no longer seek out or demand products that are less damaging to the environment." The report also pertinently observes that "some environmental claims might lead the public to believe that either the problem does not exist or that it has been solved." This could in turn undermine efforts to deal with the problems.

THE SEARCH FOR STANDARDS

To bring some order to the myriad environmental claims floating around, the attorney generals' task force is calling for states to work with the Environmental Protection Agency and the Federal Trade

Commission (FTC) to come up with "uniform national guidelines for environmental marketing claims." The task force recommends that environmental claims be as specific as possible and be supported by "reliable scientific evidence." It also proposes that manufacturers only be able to claim a product is recyclable or degradable if the advertised disposal option is currently available to consumers.

National environmental seal of approval programs are already operating in Germany, Canada, and Japan. The European Community is preparing to attach a green label to consumer products that are manufactured in the most environmentally safe method. In this country, several states are beginning to regulate environmental claims. As of 1991, California mandates that any product advertised as recycled must contain at least 10% by weight of "postconsumer" material. Rhode Island permits "recyclable" labels only if 75% of the population has access to collection systems and recycling facilities. New York State regulations place similar restrictions on the use of the term "recyclable."

Many private companies have actually expressed support for national labeling standards, in part to head off aggressive state regulations. In February 1991, a coalition of private trade organizations submitted proposed guidelines governing environmental claims to the Federal Trade Commission. The petitioners clearly hope that the FTC will use their suggestions to develop national standards for such terms as recycled, recyclable, compostable, refillable, and source-reduced. Though FTC guidelines wouldn't necessarily proscribe states' adopting more stringent standards, the manufacturers obviously wish to halt the states' regulatory momentum. Glenn Gamber of the National Food Processors Association (NFPA) told *The New York Times* that his organization wants to "take the steam out of the urge of individual states to regulate these terms." NFPA president John Cady claimed that "no company will invest in environmental marketing if it must comply with 50 conflicting state standards."

But environmental and consumer organizations are critical of the industry petition. "Industry would like to have a tool to put the brakes on state and local regulations," says Craig Merrilees, of the National Toxics Campaign. "We are vehemently opposed to that." He is concerned that FTC guidelines may be weaker than many state and local regulations; and though they don't carry the force of law, he is afraid that industry could use them to block more stringent local rules. He is particularly critical of industry's suggestion that the term "recyclable" be permitted as long as the claim specifies where recycling facilities exist. Because almost all waste is technically recyclable, and because the ability to recycle many materials is so limited, he asserts, that claim is rendered meaningless. He suggests that focusing specifically on the amount of recycled material in a product is much more important when it comes to promoting recycling.

In the absence of any government-sanctioned nationwide system of standards, two competing environmental ratings systems have sprung up. The first one to appear was the Green Cross Certification Corp. (GCCC), a nonprofit division of Scientific Certification Systems in Oakland, California, that started evaluating environmental claims in 1990. Green Cross began by focusing its efforts on evaluating the amount of recycled materials in products. To earn a Green Cross seal of approval for recycled content, a product must match or exceed 80% "of the level verified to be 'state-of-the-art,' " which is defined as the greatest percentage of recycled material that can feasibly be used in the product. To qualify for Green Cross certification for biodegradability, a soap, detergent, or cleanser must break down into "simple" substances such as carbon dioxide, water, and minerals, and must not adversely affect wildlife. By early 1991, GCCC had certified claims for more than 200 products.

Obtaining Green Cross certification is completely voluntary on the part of manufacturers. Companies that sign up with Green Cross consent to plant inspection, to paying for lab fees, and to staying abreast of the latest technology for improving the environmental compatibility of their products. It was Green Cross that pointed out the deficiencies in Willamette Industries' claims that its paper bags were recyclable and biodegradable, which led to the firm's retraction of those assertions.

When it first started giving out its certifications, Green Cross was criticized for limiting its evaluations to relatively narrow parameters, such as the percentage of recycled material in a product. Why should a product be stamped with a Green Cross, it was asked, just because it contains a certain percentage of recycled material, when its disposal or use may create pollution problems? Carl Woestendiek of the Seattle Solid Waste Utility, for example, told the *Seattle Post-Intelligencer* that a brand of compressed fireplace logs called Fire To Go received a Green Cross simply because they are made of 100% recycled materials, even though burning them produces toxic wood smoke.

Perhaps in response to criticism, Green Cross announced in January 1991 that it would broaden the criteria it uses to evaluate "environmentally friendly" products by using a "life cycle" analysis, which looks at a product's complete effect on the environment. This means evaluating everything from the raw materials that go into its manufacture to its distribution, use, and disposal. "If a product isn't the best in every single category, we'll withhold our certification," vowed GCCC president Stanley Rhodes.

Another system, called Green Seal, which is the brainchild of Earth Day founder Denis Hayes, also evaluates a product's total impact on the environment, from manufacture to disposal. Green Seal, which enjoys support from a number of environmental and consumer groups, is a nonprofit organization financed by a combination of foundation grants, private donations, and fees from manufacturers who submit their products for analysis. It plans to develop criteria for the product categories selected for evaluation and invites public participation in the process. In June 1991, Green Seal proposed an environmental standard for facial and toilet tissue. To get its stamp of approval, Green Seal suggested that, among other criteria, tissue be made of 100% wastepaper (and at least 10% postconsumer waste), contain no added dyes or fragrances, and not be de-inked with certain toxic substances.

Unfortunately, until national standards for environmental claims are developed, consumers are going to have to sort through all the hype, buzzwords, and missing information themselves. The bottom line is that consuming is *never* good for the environment, and we are not going to save the world by shopping. The manufacture, use, and ultimate disposal of consumer products eats up energy and natural resources, pollutes, and generally creates waste that must be disposed of. The purpose of purchasing and using green products is to minimize that harm. Selecting such products involves value judgments by consumers as well as considerations of convenience and price. To aid in making those decisions, manufacturers must give consumers access to truthful information, presented in as clear and understandable a manner as possible.

Resources

The following listing includes dealers in environmentally friendly goods, sourcebooks for such goods, and product-rating organizations. Our research turned up few, if any, sources in Montana and Idaho, though many food and supply stores are making efforts to stock environmentally "cleaner" products.

Catalogs

Real Goods Trading
966 Mazzoni Street
Ukiah, CA 95482
(800) 762-7325

Seventh Generation
Colchester, VT 05446
(800) 456-1177

Stores

Down to Earth
500 Olive Street
Eugene, OR 97401
(503) 344-6357

Earth Goods
6317 Roosevelt Way NE
Seattle, WA 98115
(206) 523-0977

Earth Mercantile
6345 SW Capitol Highway
Portland, OR 97201
(503) 246-4935

ECO-LOGICAL WISDOM
1705 N 45th Street
Seattle, WA 98103
(206) 548-1334

ECOLOGY HOUSE
341 SW Morrison Street
Portland, OR 97204
(503) 223-4883

IF NOT NOW...WHEN?
512 NW 21st Avenue
Portland, OR 97209
(503) 222-4471

BOOKS

THE GREEN CONSUMER
J. Elkington, J. Hailes, and J. Makower
New York: Penguin Books, 1990

THE GREEN CONSUMER SUPERMARKET GUIDE
J. Elkington, J. Hailes, and J. Makower
New York: Penguin Books, 1991

SHOPPING FOR A BETTER WORLD
Ben Corson et al.
New York: Council on Economic Priorities and Ballantine Books, 1990

ORGANIZATIONS

COUNCIL ON ECONOMIC PRIORITIES
30 Irving Place
New York, NY 10003
(800) 822-6435

Nonprofit research organization focusing on social issues and corporate responsibility. Publications include *Shopping for a Better World*, a guide to environmentally, socially, and politically correct shopping.

Green Products Resources 199

GREEN CROSS
Scientific Certification Systems Inc.
1611 Telegraph Avenue, Suite 111
Oakland, CA 94612-2113
(800) 829-1416

See description in this chapter.

GREEN SEAL INC.
PO Box 1694
Palo Alto, CA 94302
(415) 327-2200

See description in this chapter.

WASHINGTON TOXICS COALITION
4516 University Way NE
Seattle, WA 98105
(206) 632-1545

The coalition's attack is three-pronged: (1) find and develop alternatives to home, agriculture, and forestry toxics, (2) protect groundwater, and (3) reduce industrial toxics. Approach is mainly educational, publishing articles and newsletters and mobilizing communities against local environmental problems through letter-writing and planning-committee meetings. Membership organization, financial donations crucial. Newsletter, some volunteer opportunities in working with project coordinators.

APPENDIXES

STATE AGENCIES

FEDERAL AGENCIES

APPENDIX I
STATE AGENCIES

IDAHO

DEPARTMENT OF EDUCATION
650 W State Street
Boise, ID 83720
(208) 334-2281

DEPARTMENT OF LANDS
1215 W State Street
Boise, ID 83720-7000
(208) 334-0200

DIVISION OF ENVIRONMENTAL QUALITY
DEPARTMENT OF HEALTH AND WELFARE
1410 N Hilton Street
Boise, ID 83706
(208) 334-5840

FISH AND GAME DEPARTMENT
600 S Walnut Street, Box 25
Boise, ID 83707
(208) 334-3700

THE HONORABLE CECIL ANDRUS
Office of the Governor
State House
Boise, ID 83720
(208) 334-2100

MONTANA

DEPARTMENT OF FISH,
WILDLIFE AND PARKS
1420 E Sixth Avenue
Helena, MT 59620
(406) 444-2535

**DEPARTMENT OF HEALTH
AND ENVIRONMENTAL SCIENCES**
*Cogswell Building
1400 Broadway
Helena, MT 59620
(406) 444-2544*

DEPARTMENT OF STATE LANDS
*1625 11th Avenue
Capitol Station
Helena, MT 59620
(406) 444-2074*

THE HONORABLE STAN STEPHENS
*Capitol Station
Helena, MT 59620
(406) 444-3111*

OFFICE OF PUBLIC INSTRUCTION
*Capitol Building
Helena, MT 59620
(406) 444-3680*

OREGON

DEPARTMENT OF EDUCATION
*700 Pringle Parkway SE
Salem, OR 97310
(503) 373-7898*

DEPARTMENT OF ENVIRONMENTAL QUALITY
*811 SW Sixth Avenue
Portland, OR 97204
(503) 229-5696*

DEPARTMENT OF FISH AND WILDLIFE
*PO Box 59
Portland, OR 97207
(503) 229-5403*

DEPARTMENT OF FORESTRY
2600 State Street
Salem, OR 97310
(503) 378-2560

THE HONORABLE BARBARA ROBERTS
State Capitol, Room 254
Salem, OR 97310
(503) 378-3111

WASHINGTON

DEPARTMENT OF ECOLOGY
Abbot Raphael Hall
Mail Stop PV-11
Olympia, WA 98504-8711
(206) 459-6000

DEPARTMENT OF FISHERIES
[Saltwater fish, salmon, shellfish]
115 General Administration Building
Olympia, WA 98504
(206) 753-6600

DEPARTMENT OF NATURAL RESOURCES
John A. Cherbert Building, Room 201
Olympia, WA 98504
(206) 753-5327

DEPARTMENT OF WILDLIFE
[Freshwater fish, trout, game fish]
600 Capitol Way N
Olympia, WA 98501-1091
(206) 753-5700

THE HONORABLE BOOTH GARDNER
Office of the Governor
Legislative Building
Olympia, WA 98504
(206) 753-6780

OFFICE OF THE STATE SUPERINTENDENT
OF PUBLIC INSTRUCTION
Old Capitol Building
Olympia, WA 98504
(206) 753-6757

PUGET SOUND WATER QUALITY AUTHORITY
Mail Stop PV-15
Olympia, WA 98504
(206) 493-9300

APPENDIX II
FEDERAL AGENCIES

BUREAU OF LAND MANAGEMENT

Idaho State Director's Office
3380 Americana Terrace
Boise, ID 83706
(208) 384-3001

Montana State Director's Office
PO Box 36800
Billings, MT 59107
(406) 255-2904

Oregon State Office
PO Box 2965
Portland, OR 97208
(503) 280-7024

Washington State Office
(see Oregon)

NATIONAL MARINE FISHERIES SERVICE

Northwest Region
NOAA
7600 Sand Point Way NE
Bin C15700 Building 1
Seattle, WA 98115
(206) 526-6150

NATIONAL PARK SERVICE
(Each park also has its own office.)

Pacific Northwest Office
[ID, OR, WA]
83 S King Street
Seattle, WA 98104
(206) 442-5565

Rocky Mountain Regional Office
[CO, MT, ND, SD, UT, WY]
PO Box 25287
Denver, CO 80225
(303) 969-2500

UNITED STATES ENVIRONMENTAL PROTECTION AGENCY

Region 8
[CO, MT, ND, SD, UT, WY]
999 18th Street, Suite 500
Denver, CO 80202-2405
(303) 293-1603

Region 10
[AK, ID, OR, WA]
1200 Sixth Avenue
Seattle, WA 98101
(206) 553-1200

UNITED STATES FOREST SERVICE

Region 1
[Northern ID, MT, ND National Grasslands]
PO Box 7669
Missoula, MT 59807
(406) 329-3511

Region 4
[Southern ID, NV, UT, Western WY]
324 25th Street
Ogden, UT 84401
(801) 625-5347

Region 6
[OR, WA]
PO Box 3623
Portland, OR 97208
(503) 326-2877

UNITED STATES FISH AND WILDLIFE SERVICE

Pacific Northwest Regional Office
911 NE 11th Avenue
Portland, OR 97232-4181
(503) 231-6121

Rocky Mountain Regional Office
PO Box 25486
Denver Federal Center
Denver, CO 80225
(303) 236-7920

ACKNOWLEDGMENTS

It is impossible to acknowledge everyone who has contributed to this effort, but several people deserve to be singled out: my editor, Anne Depue, who always listened to what I had to say; Chad Haight, for backing the project; Emily Hall, who researched and wrote most of the environmental listings and chased down a lot of information; Barry Foy, who smoothed out some rough language; Ina Chang, who offered invaluable editing and organizational suggestions; Todd Campbell, who gave freely of his time to provide moral support and editorial suggestions; and Mark Terry, who reviewed the manuscript and provided valuable observations. Many individuals took a lot of time out to lend their expertise to the project; I'd particularly like to thank Jane Ceraso, Daniel Jack Chasan, Jules James, and Kent Swigard. Special appreciation goes to Sally Anderson, who helped with her considerable editorial and grammatical skills, and who was understanding of deadline pressure.

ABOUT THE AUTHOR

Jonathan King is a Seattle writer and was formerly news editor of *Seattle Weekly* and a reporter for *ABC World News Tonight*'s investigative unit. He has written on national security, environmental, and health issues for a variety of national publications, and is author of *Troubled Water: The Poisoning of America's Drinking Water (Rodale Press, 1985)*.

DID YOU ENJOY THIS BOOK?

Sasquatch Books publishes books and guides related to the Pacific Northwest. Our books are available at bookstores and other retail outlets throughout the region. Here is a selection of our current titles:

GUIDEBOOKS

Northwest Best Places
Restaurants, Lodgings, and Touring in Oregon, Washington, and British Columbia
David Brewster and Stephanie Irving

Portland Best Places
A Discriminating Guide to Portland's Restaurants, Lodgings, Shopping, Nightlife, Arts, Sights, Outings, and Annual Events
Stephanie Irving

Seattle Best Places
A Discriminating Guide to Seattle's Restaurants, Lodgings, Shopping, Nightlife, Arts, Sights, Outings, and Annual Events
David Brewster and Stephanie Irving

Seattle Cheap Eats
300 Terrific Bargain Eateries
Kathryn Robinson and Stephanie Irving

Seattle Survival Guide
The Essential Handbook for City Living
Theresa Morrow

Washington Homes
Buying, Selling, and Investing in Seattle and Statewide Real Estate
Jim Stacey

Field Guide to the Bald Eagle
The Audubon Society

Field Guide to the Gray Whale
American Cetacean Society

Field Guide to the Orca
The Oceanic Society

GARDENING

The Border in Bloom
A Northwest Garden Through the Seasons
Ann Lovejoy

Gardening Under Cover
A Northwest Guide to Solar Greenhouses, Cold Frames, and Cloches
William Head

Growing Vegetables West of the Cascades
Steve Solomon's Complete Guide to Natural Gardening
3rd edition

Trees of Seattle
The Complete Tree-finder's Guide to 740 Varieties
Arthur Lee Jacobson

Winter Gardening in the Maritime Northwest
Cool Season Crops for the Year-Round Gardener
Binda Colebrook, 3rd edition

The Year in Bloom
Gardening for All Seasons in the Pacific Northwest
Ann Lovejoy

The Territorial Seed Company Garden Cookbook
Homegrown Recipes for Every Season
Edited by *Lane Morgan*

Winter Harvest Cookbook
How to Select and Prepare Fresh Seasonal Produce All Winter Long
Lane Morgan

To receive a Sasquatch Books catalog, or to inquire about ordering our books by phone or mail, please contact us at the address below.

SASQUATCH BOOKS   1931 Second Avenue
Seattle, Washington 98101   (206) 441-5555